MANAGEMENT STRATEGIES FOR WATER USE EFFICIENCY AND MICRO IRRIGATED CROPS

Principles, Practices, and Performance

Innovations and Challenges in Micro Irrigation

MANAGEMENT STRATEGIES FOR WATER USE EFFICIENCY AND MICRO IRRIGATED CROPS

Principles, Practices, and Performance

Edited by

Megh R. Goyal, PhD, P.E.
B. J. Pandian, PhD

(in Cooperation with Water Technology Center at TNAU, Coimbatore)

Apple Academic Press Inc. | Apple Academic Press Inc.
3333 Mistwell Crescent | 1265 Goldenrod Circle NE
Oakville, ON L6L 0A2 | Palm Bay, Florida 32905
Canada USA | USA

© 2019 by Apple Academic Press, Inc.

First issued in paperback 2021

Exclusive worldwide distribution by CRC Press, a member of Taylor & Francis Group
No claim to original U.S. Government works

ISBN 13: 978-1-77463-467-7 (pbk)
ISBN 13: 978-1-77188-791-5 (hbk)

CIP data on file with Canada Library and Archives

CIP data on file with US Library of Congress

Apple Academic Press also publishes its books in a variety of electronic formats. Some content that appears in print may not be available in electronic format. For information about Apple Academic Press products, visit our website at **www.appleacademicpress.com** and the CRC Press website at **www.crcpress.com**

OTHER BOOKS ON MICRO IRRIGATION TECHNOLOGY BY APPLE ACADEMIC PRESS, INC.

Management of Drip/Trickle or Micro Irrigation
Megh R. Goyal, PhD, PE, Senior Editor-in-Chief

Evapotranspiration: Principles and Applications for Water Management
Megh R. Goyal, PhD, PE, and Eric W. Harmsen, Editors

Book Series: Research Advances in Sustainable Micro Irrigation
Senior Editor-in-Chief: Megh R. Goyal, PhD, PE

Volume 1: Sustainable Micro Irrigation: Principles and Practices
Volume 2: Sustainable Practices in Surface and Subsurface Micro Irrigation
Volume 3: Sustainable Micro Irrigation Management for Trees and Vines
Volume 4: Management, Performance, and Applications of Micro Irrigation Systems
Volume 5: Applications of Furrow and Micro Irrigation in Arid and Semi-Arid Regions
Volume 6: Best Management Practices for Drip Irrigated Crops
Volume 7: Closed Circuit Micro Irrigation Design: Theory and Applications
Volume 8: Wastewater Management for Irrigation: Principles and Practices
Volume 9: Water and Fertigation Management in Micro Irrigation
Volume 10: Innovation in Micro Irrigation Technology

Book Series: Innovations and Challenges in Micro Irrigation
Senior Editor-in-Chief: Megh R. Goyal, PhD, PE

- Engineering Interventions in Sustainable Trickle Irrigation
- Micro Irrigation Engineering for Horticultural Crops: Policy Options, Scheduling and Design

- Micro Irrigation Management: Technological Advances and Their Applications
- Micro Irrigation Scheduling and Practices
- Performance Evaluation of Micro Irrigation Management: Principles and Practices
- Potential Use of Solar Energy and Emerging Technologies in Micro Irrigation
- Principles and Management of Clogging in Micro Irrigation
- Sustainable Micro Irrigation Design Systems for Agricultural Crops: Methods and Practices
- Management Strategies for Water Use Efficiency and Micro Irrigated Crops: Principles, Practices, and Performance

ABOUT THE SENIOR EDITOR-IN-CHIEF

Megh R. Goyal, PhD

Retired Professor in Agricultural and Biomedical Engineering, University of Puerto Rico, Mayaguez Campus; Senior Acquisitions Editor, Biomedical Engineering and Agricultural Science, Apple Academic Press, Inc.

Megh R. Goyal, PhD, PE, is a Retired Professor in Agricultural and Biomedical Engineering from the General Engineering Department in the College of Engineering at the University of Puerto Rico–Mayaguez Campus; and Senior Acquisitions Editor and Senior Technical Editor-in-Chief in Agriculture and Biomedical Engineering for Apple Academic Press, Inc. He has worked as a Soil Conservation Inspector and as a Research Assistant at Haryana Agricultural University and Ohio State University.

During his professional career of 45 years, Dr. Goyal has received many prestigious awards and honors. He was the first agricultural engineer to receive the professional license in Agricultural Engineering in 1986 from the College of Engineers and Surveyors of Puerto Rico. In 2005, he was proclaimed as "Father of Irrigation Engineering in Puerto Rico for the Twentieth Century" by the American Society of Agricultural and Biological Engineers (ASABE), Puerto Rico Section, for his pioneering work on micro irrigation, evapotranspiration, agroclimatology, and soil and water engineering. The Water Technology Centre of Tamil Nadu Agricultural University in Coimbatore, India, recognized Dr. Goyal as one of the experts "who rendered meritorious service for the development of micro irrigation sector in India" by bestowing the Award of Outstanding Contribution in Micro Irrigation. This award was presented to Dr. Goyal during the inaugural session of the National Congress on "New Challenges and Advances in Sustainable Micro Irrigation" on March 1, 2017, held at Tamil Nadu Agricultural University. Dr. Goyal received the Netafim Award for Advancements in Microirrigation 2018 from the American

Society of Agricultural Engineers at the ASABE International Meeting in August 2018.

A prolific author and editor, he has written more than 200 journal articles and textbooks and has edited over 55 books. He is the editor of three book series published by Apple Academic Press: Innovations in Agricultural & Biological Engineering, Innovations and Challenges in Micro Irrigation, and Research Advances in Sustainable Micro Irrigation. He is also instrumental in the development of the new book series Innovations in Plant Science for Better Health: From Soil to Fork.

Dr. Goyal received his BSc degree in engineering from Punjab Agricultural University, Ludhiana, India; his MSc and PhD degrees from Ohio State University, Columbus; and his Master of Divinity degree from Puerto Rico Evangelical Seminary, Hato Rey, Puerto Rico, USA.

ABOUT THE EDITOR

B. J. Pandian, PhD
(Organizing Chairman of NCMI – 2017)
Director, Water Technology Centre, Tamil Nadu Agricultural University in Coimbatore, India

B. J. Pandian, PhD, is presently working as Director of the Water Technology Centre, Tamil Nadu Agricultural University in Coimbatore, India. He has served the university for 34 years, with 16 years as a professor in teaching, research, extension, and administrative aspects. He has taught undergraduate and postgraduate agronomy courses and brought out course books and manuals.

Dr. Pandian was instrumental in developing and promoting the System of Rice Intensification (SRI) and the Sustainable Sugarcane Initiative (SSI) in Tamil Nadu, India. He has also worked on the promotion of drip fertigation in horticulture and agriculture crops, especially in rice crops. He handled many projects for the Indian Council of Agricultural Research, private agencies, and the World Bank. He successfully implemented the Tamil Nadu Irrigated Agricultural Modernization and Water Bodies Restoration and Management Project (TN-IAMWARM), funded by the World Bank for the last seven years, which made a remarkable impact on improving water-saving technologies in Tamil Nadu.

He has participated in many national and international seminars and has organized several himself, such as two international conferences, on "Drip fertigation in rice" and "Exploring the emergence and spread of the SRI in India," and several national conferences, such as the "National Groundwater Conference on Problems, Challenges and Management of Groundwater in Agriculture," "Sustainable Sugarcane Initiative (SSI)," and "Wasteland Development." He was awarded for his outstanding extension work by the Government of Tamil Nadu in 1983.

Dr. Pandian has published 15 international and 95 national research papers, 32 technical books and 13 course books and more than 100 popular articles. He is a member of many agriculture and water resources organizations.

He graduated from Tamil Nadu Agricultural University in Coimbatore, India, and completed his master and doctoral degrees in the field of agriculture with specialization in agronomy.

CONTENTS

CONTRIBUTORS

S. Alagudurai
Krishi Vigyan Kendra, Veterinary College and Research Institute Campus, Namakkal – 637002, Tamil Nadu – India, Mobile: +91-9786679700, E-mail: jaialagu@gmail.com

Sunitha Sarojini Amma
ICAR-Central Tuber Crops Research Institute, Sreekariyam, Aakkulam Road, Near Loyola School, Trivandrum – 695017, Kerala, India, Mobile: +91-9446396026,
E-mail: sunitharajan1@rediffmail.com

K. Arthi
Department of Agricultural Economics, Tamil Nadu Agricultural University, Lawley Road, Coimbatore – 641003, Tamil Nadu, India, Mobile: +91-8344636993, E-mail: arthi709@gmail.com

Gurusamy Arumugam
Department of Agronomy, Agricultural College and Research Institute, Tamil Nadu Agricultural University, Kudumiyanmalai – 622104, Tamil Nadu, India, Mobile: +91-9443973236,
E-mail: guruwms2009@gmail.com

J. Auxcilia
Department of Fruit Crops, Horticultural College and Research Institute (HC & RI), Tamil Nadu Agricultural University, Lawley Road, Coimbatore – 641003, India, E-mail: aux1@rediffmail.com

M. Babu
Tamil Nadu Veterinary and Animal Sciences University, Cosy Nest, 6.Carmel Garden, Ramalinga Nagar West Extension, Tiruchirapalli-620003, India, Mobile: +91-9443924055,
E-mail: drbabum@gmail.com

R. Balasubramanian
Department of Market Extension, Tamil Nadu Agricultural University, Lawley Road, Coimbatore   641003, Tamil Nadu, India, Mobile: +91-9965572120, E-mail: rubalu@gmail.com

R. Chandrasekaran
Rice Research Station, Tamil Nadu Agricultural University, Tirurkuppam, Tirur – 602025, Thiruvallore District, Tamil Nadu, India, Mobile: +91-9486201357, E-mail: chandru_tnau@yahoo.co.in

C. Chinnusamy
Department of Agronomy, Tamil Nadu Agricultural University, Coimbatore – 641003, Tamil Nadu, India, Mobile: +91-9443721575, E-mail: chinnusamyc@gmail.com

M. Deiveegan
Department of Agronomy, Agricultural College & Research Institute (AC & RI), Tamil Nadu Agricultural University (TNAU), Coimbatore – 641003, Tamil Nadu, India, Mobile: +91-9843205593, E-mail: devedeva07@gmail.com

P. Devasenapathy
Department of Agronomy, Tamil Nadu Agricultural University, Coimbatore – 641003, Tamil Nadu, India, Mobile: +91-9443319381, E-mail: devasenapathy@gmail.com

Pradip Dey
ICAR – Indian Institute of Soil Science (IISS), Nabi Bagh, Berasia Road, Bhopal – 462038, Madhya Pradesh, India, Mobile: +91-9425608219, E-mail: pradipdey@yahoo.com

K. Divya
Department of Agricultural and Rural Management, Tamil Nadu Agricultural University, Lawley Road, Coimbatore – 641003, Tamil Nadu, India, Mobile: +91-9443899683, E-mail: divyatnau@gmail.com

V. K. Duraisamy
Directorate of Agricultural Business Development, Tamil Nadu Agricultural University, Coimbatore 641003, Tamil Nadu, India, Mobile: +91-9443853473, E-mail: vkduraisamy@yahoo.com

James George
AICRP (All India Co-ordinated Research Project on Tuber Crops), ICAR-Central Tuber Crops Research Institute, Sreekariyam, Aakkulam Road, Near Loyola School, Trivandrum- 695017, Kerala, India, Mobile: +91-9447111289, E-mail: jgkarott@gmail.com

H. Gopi
Post Graduate Research Institute in Animal Sciences, Tamil Nadu Veterinary and Animal Sciences University, Kattupakkam – 603203, Tamil Nadu, Mobile: +91-9840036268, E-mail: drhgopi@gmail.com

Megh R. Goyal
University of Puerto Rico – Mayaguez Campus, Senior Editor-in-Chief, Apple Academic Press Inc., Mailing Address: P.O. Box 86, Rincon – Puerto Rico, USA, E-mail: goyalmegh@gmail.com

V. M. Abdul Hakkim
Department of Land and Water Resources and Conservation Engineering, Kerala College of Agricultural Engineering & Technology (KCAET), Kerala Agricultural University, Tavanur (P.O.) – 679573, Malappuram (District), Kerala, Mobile: +91-9446279626, E-mail: abdulhakkim.vm@kau.in

M. Jawaharlal
Horticultural College and Research Institute, Tamil Nadu Agricultural University (TNAU), Coimbatore – 641003, Tamil Nadu, India, Mobile: +91-422-6611371, E-mail: deanhortcbe@tnau.ac.in

R. Jeyajothi
Department of Agronomy, Agricultural College & Research Institute (AC & RI), Tamil Nadu Agricultural University (TNAU), Coimbatore – 641003, Tamil Nadu, India, Mobile: +91-7598479852, E-mail: jeyajothi.rose@gmail.com

J. Kabariel
Adhiyamaan College of Agriculture and Research, Athimugam Village, Shoolagiri TK, Hosur – 635105, Krishanagiri District, Tamil Nadu, India, Mobile: +91-9943545589, E-mail: kabara2y@yahoo.co.in

M. Kannan
Adhiparasakthi Horticultural College, TNAU, G. B. Nagar, Kalavai 63250, Vellore District, Tamil Nadu, India

V. Ramesh Saravana Kumar
Centre for Animal Production Studies, Tamil Nadu Veterinary Animal Sciences University, Chennai – 600051, India, Mobile: +91-9443544351, E-mail: rameshsaravanakumar.v@tanuvas.ac.in

Anbarasu Mariyappillai
Department of Agronomy, Agricultural College and Research Institute, Tamil Nadu Agricultural University, Madurai – 625104, Tamil Nadu, India, Mobile: +91-9786185701,
E-mail: manbarasu102@gmail.com

M. Murugan
Department of Livestock Production and Management, Veterinary College and Research Institute, Tirunelveli – 627358, Tamil Nadu, Mobile: +91-9444688273, E-mail: lpmmurugan@gmail.com

R. Murugeswari
Institute of Animal Nutrition, Tamil Nadu Veterinary Animal Sciences University, Kattupakkam – 603203, TN – India, Mobile: +91-9047427561,
E-mail: murugeswari.r@tanuvas.ac.in

P. Muthulakshmi
Department of Fruit Crops, Horticultural College and Research Institute (HC & RI), Tamil Nadu Agricultural University, Lawley Road, Coimbatore – 641003, India,
E-mail: muthupathology@gmail.com

V. S. Mynavathi
Institute of Animal Nutrition, Tamil Nadu Veterinary Animal Sciences University, Kattupakkam – 603203, TN – India, Mobile: +91-9942965516, E-mail: mynagri@gmail.com

R. Nageswari
Sugarcane Research Station, Tamil Nadu Agricultural University, Sirugamani, Trichy – 639115, Tamil Nadu, India, Mobile: +91-7502840470, E-mail: oryzanagtn@gmail.com

S. K. Natarajan
Agricultural Research Station, Tamil Nadu Agricultural University, Bhavanisagar – 638451, Tamil Nadu, India, Mobile: +91-9626919760, E-mail: kandunats@gmail.com

B. J. Pandian
Water Technology Center, Tamil Nadu Agricultural University (TNAU), Coimbatore – 641003, Tamil Nadu, India, Mobile: +91-9443286711, E-mail: directorwtc@tnau.ac.in

Prakash Patil
ICAR – Indian Institute of Horticultural Research, Hesaraghatta Lake Post, Bengaluru – 560089, Karnataka, Email: pcfruits@gmail.com

S. Pazhanivelan
Department of Remote Sensing and GIS, Agricultural College & Research Institute (AC & RI), Tamil Nadu Agricultural University (TNAU), Coimbatore – 641003, Tamil Nadu, India,
Mobile: +91-9047599446, E-mail: pazhanivelans@gmail.com

K. R. Pushpanathan
Instructional Livestock Farm Complex, Veterinary College and Research Institute at Salem – Namakkal, Trichy Road, Thillaipuram, Namakkal – 637002, Tamil Nadu, India,
Mobile: +91–9047949976, E-mail: pushpanathanr4@gmail.com

D. David Rajasekar
Agricultural Economics, Agricultural College and Research Institute (AC&RI), Tamil Nadu Agricultural University (TNAU), Madurai – 625104, Tamil Nadu, India, Mobile: +91-9865144857,
E-mail: david250760@rediffmail.com

Indirani Ramesh
Department of Soil and Environment, Agricultural College and Research Institute, Tamil Nadu Agricultural University, Madurai – 625104, Tamil Nadu, India, Mobile: +91-9443714971, E-mail: indirani_ramesh@yahoo.co.in

Santhi Rangasamy
Department of Soil Science and Agricultural Chemistry, Tamil Nadu Agricultural University, Coimbatore – 641003, Tamil Nadu, India, Mobile: +91-9865092150, E-mail: santhitnau@yahoo.co.in

S. Ravichandran
Department of Agronomy, Agricultural College & Research Institute (AC & RI), Tamil Nadu Agricultural University (TNAU), Coimbatore – 641003, Tamil Nadu, India, Mobile: +91 8793062079, E-mail: ravichandran@rubberboard.org.in

A. Revathy
Agricultural Economics, Agricultural College and Research Institute (AC&RI), Tamil Nadu Agricultural University (TNAU), Coimbatore – 641003, Tamil Nadu, India, Mobile: +91-9790242666, E-mail: revathyg11@gmail.com

K. S. Sangeetha
Department of Plantation Crops and Spices, College of Horticulture, Kerala Agricultural University, Thrissur – 680656, India, Mobile: +91-9497246229, E-mail: sangy.666@gmail.com

T. Saranraj
Department of Agronomy, Tamil Nadu Agricultural University, Coimbatore – 641003, Tamil Nadu, India, Mobile: +91-7892451675, E-mail: tsaranrajagronomy@gmail.com

V. Saravanakumar
Department of Agricultural Economics, Tamil Nadu Agricultural University (TNAU), Lawley Road, Coimbatore – 641003, Tamil Nadu, India, Mobile: +91-9442267934, E-mail: sharanu2k@gmail.com

Selvaraj Selvakumar
Department of Soil and Water Conservation Engineering, Tamil Nadu Agricultural University, Krishi Vigyan Kendra, Arrupukottai – 641003, Tamil Nadu, India, Mobile: +91-9487626413, E-mail: engineeringselva@yahoo.co.in

S. Somasundaram
Department of Agronomy, Anbil Dharmalingam Agricultural College and Research Institute, Tamil Nadu Agricultural University (TNAU), Dindugal Main Road, Muthukkulam, Navalur Kottapattu – 620009, Tiruchirappalli, Tamil Nadu, India, Mobile: +91-9965948419, E-mail: rainfed@yahoo.com

K. Soorianathasundaram
Department of Fruit Crops, Horticultural College and Research Institute (HC & RI), Tamil Nadu Agricultural University, Lawley Road, Coimbatore – 641003, India, E-mail: sooria@tnau.ac.in

Praveena Katharine Stephen
Institute of Agricultural Engineering College & Research Institute (IAECRC), Kumulur – 621712, Trichy District, Tamil Nadu, India, Mobile: +91-9585736885, E-mail: praveenakate@rediffmail.com

K. B. Sujatha
Department of Fruit Crops, Horticultural College and Research Institute (HC & RI), Tamil Nadu Agricultural University, Lawley Road, Coimbatore – 641003, India, E-mail: kb.sujatha@rediffmail.com

Anjaly C. Sunny

Department of Land and Water Resources and Conservation Engineering, College of Agricultural Engineering & Technology (KCAET), Kerala Agricultural University, Tavanur (P.O.) – 679573, Malappuram (District), Kerala, Mobile: +91-8281779569, E-mail: anjuminju@gmail.com

J. Suresh

Department of Spices and Plantation Crops, Horticultural College and Research Institute (HC & RI), Tamil Nadu Agricultural University, Lawley Road, Coimbatore -641003, India, Mobile: +91-9489056732, E-mail: spices@tnau.ac.in

K. S. Usharani

Plant Breeding and Genetics, Agricultural Research Station, Tamil Nadu Agricultural University, Bhavanisagar 638451, Tamil Nadu, India, Mobile: +91-9788479444, E-mail: usharaniagri@gmail.com

K. Vaiyapuri

Department of Agronomy, Tamil Nadu Agricultural University (TNAU), Coimbatore – 641003, Tamil Nadu, India, Mobile: +91-9488214505, E-mail: vaistnau@gmail.com

C. Valli

Institute of Animal Nutrition, Tamil Nadu Veterinary Animal Sciences University, Kattupakkam – 603203, TN – India, Mobile: +91-9840671046, E-mail: valli.c@tanuvas.ac.in

R. M. Vijayakumar

Department of Fruit Crops, Horticultural College and Research Institute (HC & RI), Tamil Nadu Agricultural University, Lawley Road, Coimbatore – 641003, India, E-mail: fruits@tnau.ac.in

ABBREVIATIONS

ADC	analog digital converters
AE	agronomic efficiency
AICRP-STCR	All India Coordinated Research Project for Investigations on Soil Test Crop Response Correlation
ANOVA	analysis of variance
BCR	benefit cost ratio
CACP	Commission on Agricultural Costs and Prices
CARDS	Centre for Agricultural and Rural Development Studies
CCS	commercial cane sugar
CD	critical difference
CEC	cation exchange capacity
CEY	castor equivalent yield
CF	conventional fertilizer
CGR	crop growth rate
CPE	cumulative pan evaporation
CPG	crop production guide
CRD	completely randomized design
DAI	days after irrigation
DAP	days after planting
DAS	days after sowing
DCD	dicyandiamide
DF	drip fertigation
DI	drip irrigation
DMI	drip method of irrigation
DMP	dry matter production
dS m^{-1}	Deci Siemen per meter
DSR	direct seeded rice
DTPA	diethylene triamine penta acetic acid
DW	dry weights
EC	electrical conductivity
ECA	class A pan evaporation
ER	effective rainfall
ER	evaporation replenishment

ET	evapotranspiration
Evap	evaporation
FIM	furrow irrigation mulching
FIN	furrow irrigation non-mulching
FIP	fertilizer injection pump
FK2O	fertilizer potassium
FMI	furrow method of irrigation
FN	fertilizer nitrogen
FP2O5	fertilizer phosphorus
FPE	fraction of pan evaporation
FRBD	fully randomized block design
FUE	fertilizer use efficiency
FYM	farmyard manure
GCRPSs	ground cover rice production systems
GMI	global methane initiative
GOI	Government of India
HI	harvest index
HP	horse power
HT	harvest stage
HW	hand weeding
IARI	Indian Agricultural Research Institute
IR	irrigation rate
IW/CPE	irrigation water/cumulative pan evaporation
Kc	crop coefficient
K_D	distribution rate coefficient
kg/ha	kilogram per hectare
K_{HU}	hydrolysis rate constant
K_N	nitrification rate constant
KNO_3	potassium nitrate
Kp	crop factor
Kp	pan coefficient
K_V	volatilization coefficient
LAI	leaf area index
LAm	mean leaf area
LER	land equivalent ratio
MAP	mono ammonium phosphate
MAP	months after planting
MCU	micro controller unit

MGNREGS	Mahatma Gandhi National Rural Employment Guarantee Scheme
MOP	muriate of potash
MSL	mean sea level
$NH_2 - N$	urea $-$ N
$NH_4 - N$	ammoniacal $-$ N
$NH_4H_2PO_4$	mono-ammonium phosphate
NH_4NO_3	ammonium nitrate
NMC	number of millable canes
NN	neutral normal
$NO_3 - N$	nitrate $-$ N
NPK	nutrient uptake
NPKS	nitrogen, phosphorus, potassium, and sulfur
NPV	net present value
NR	nutrient requirement
NUE	nitrogen use efficiency
PE	pan evaporation
PGRIAS	Post Graduate Research Institute in Animal Sciences
pH	negative logarithm of hydrogen ion concentration
PMA	phenyl mercuric acetate
PPC	plant protection chemicals
RBD	randomized block design
RD	recommended dose
RD	root density
RDF	recommended dose of fertilizer
Re	effective rainfall
RGR	relative growth rate
RR	response ratio
RWC	relative water content
SCW	single cane weight
SF	straight fertilizers
SI	supplementary irrigation
SLW	specific leaf weight
SOP	sulphate of potash
SRI	system of rice intensification
SSDF	subsurface drip fertigation
SSDI	subsurface drip irrigation
SSI	sustainable sugarcane initiative

SSP	single super phosphate
STCR-IPNS	soil test crop response based integrated plant nutrition system
TDR	time domain reflectometer
TFC	total fixed cost
Tmax	temperature maximum
Tmin	temperature minimum
TNAU	Tamil Nadu Agricultural University
TSS	total soluble solids
TVC	total variable cost
TW	turgid weight
UAS (B)	University of Agriculture, Bangalore
WHC	water holding capacity
Wp	wetting percentage
WR	water requirement
WSF	water soluble fertilizer
WUE	water use efficiency

FOREWORD BY K. RAMASAMY

You must be the change you wish to see in the world.

— K. Ramasamy

The agriculture (irrigation) sector, which currently consumes over 80% of the available water in India, continues to be the major water-consuming sector due to the intensification of agriculture. One of the main reasons for the low coverage of irrigation is the predominant use of flood (conventional) method of irrigation and the water use efficiency under flooding is estimated to be only 35 to 40% because of huge conveyance and distribution losses.

One of the demand management strategies introduced relatively recently to control water consumption in Indian agriculture is micro irrigation (MI), which includes mainly drip irrigation method (DIM) and sprinkler irrigation method (SIM). The conveyance and distribution losses are reduced to a minimum, which result in higher water use efficiency under MI. The development of drip irrigation and sprinkler irrigation in India indicates that about 80 crops (both narrow and widely spaced crops) can be grown under micro irrigation. There are several research reports available on the development of sustainable technologies in micro irrigation that should be evaluated and promoted among the farming community.

The field of micro irrigation is interdisciplinary, as it requires knowledge of biologists, physicists, agricultural scientists, and engineers. There is an urgent need to explore and investigate the current shortcomings and challenges and to discuss the advances made in micro irrigation.

Considering the importance of micro irrigation, the Water Technology Centre of Tamil Nadu Agricultural University appropriately organized a National Congress on "New Challenges and Advances in Sustainable Micro Irrigation" during March 1–3 of 2017. Several irrigation experts were invited to share their valuable thoughts with the participants. I congratulate all the scientists of the Water Technology Centre and Dr. Megh R. Goyal for organizing this national event at the appropriate time and for their efforts in compiling the invited and contributory research papers in a volume that will be highly useful for students, researchers, and policymakers.

I wish the team of scientists of the Water Technology Centre and editors of this book volume a grand success, and once again congratulate them for their sincere efforts in bringing out this valuable publication under the leadership of Dr. Megh R. Goyal and Dr. B. J. Pandian.

K. Ramasamy, PhD
Former Vice-Chancellor, Tamil Nadu Agricultural
University, Coimbatore, 641003,
Tamil Nadu, India,
Tel.: +91-422-6611251
E-mail: vctnau@tnau.ac.in

PREFACE 1

Inadequate technical training is a serious chronic cause of failure of micro irrigation systems.
However, if the irrigator uses the expertise of professionals for consultancy, he can live a full productive and joyful life.
Giving back is very important to me, as it defines who I am.
I am an ordinary irrigation expert, as I still live like the reader.
I just can't see you, but I can enjoy that you have read my books on micro irrigation.
God bless you as you browse through my books that have been prepared for you only.
I can assure you that drip irrigation systems can potentially provide a high uniformity coefficient and distribution efficiency, only if the system is properly serviced and maintained.

— Megh R. Goyal, Drip Man

My vision for micro irrigation technology has been expanding every day and globally. After my first textbook, *Management of Drip/Trickle or Micro irrigation* and response from international readers, Apple Academic Press has published for the world community the ten-volume series on *Research Advances in Sustainable Micro irrigation*, edited by me. The current book volume is published under the book series, *Innovations and Challenges in Micro irrigation*. Both book series are musts for those interested in irrigation planning and management, namely, researchers, scientists, educators, and students.

It has been my unforgettable and fruitful experience to be part of the organizing committee for the National Congress on *"New Challenges and Advances in Sustainable Micro Irrigation"* at Tamil Nadu Agricultural University (TNAU) – Coimbatore, India, during March 1–3 of 2017. Our host, the Water Technology Centre, has impacted and touched the hearts of my wife (Subhadra Goyal) and me with their hospitality and fine detail so that we both would feel to be part of the family of TNAU. This is 100% true. Speakers at the Congress taught me the length, breadth, height, and depth of micro irrigation.

In August of 2016, I was informed by Dr. B. J. Pandian, Director of the Water Technology Centre at TNAU, that they were to hold this conference, and it was my honor to be a guest speaker at this Congress. In 2017, Vice Chancellor Dr. K. Ramasamy announced the founding to the Micro irrigation Centre of Excellence (*MICE*) under Water Technology Centre at TNAU, which is testimony of the importance of drip irrigation to reduce water scarcity. The proposal for *MICE* by me and the Chancellor has been approved for funding by the Government of Tamil Nadu, who will soon issue the official announcement. The *MICE* is the first center of its kind worldwide.

This book volume is a compilation of selected presentations at this Congress. The editors of this book volume decided to supplement the volume with additional chapters from outside the Congress to benefit the readers.

The contributions by the cooperating authors to this book have been most valuable in the compilation of this volume. Their names are mentioned in each chapter and in the list of contributors. This book would not have been written without the valuable cooperation of micro irrigation experts, and these investigators are renowned scientists who have worked in the field of micro irrigation throughout their professional careers.

The goal of this book is to guide the world science community on the application of micro irrigation technology in agricultural crops.

I express my deep admiration to my wife, Subhadra Devi Goyal, for her understanding and collaboration during the preparation of this book. This book volume was prepared during the week when the American Society of Agricultural & Biological Engineers (ASABE) bestowed on me the 2018 Netafim Award in Advancement in Micro Irrigation on August 1 of 2018 (my birthday) in Detroit, Michigan. This recognition was based on my work on drip/trickle or micro irrigation since 1979. I owe this award to readers of my two book series.

As an educator, there is a piece of advice to one and all in the world: *"Permit that our Almighty God, our Creator, excellent Teacher, and Micro irrigation Designer, irrigate our life with His Grace of rain trickle by trickle, because our life must continue trickling on . . ."*

—**Megh R. Goyal, PhD, PE**
Senior Editor-in-Chief

PREFACE 2

Water is considered as the most critical resource for sustainable development in most countries. It is essential not only for agriculture, industry, and economic growth, but it also is the most important component of the environment, with significant impact on health and nature conservation.

The global irrigated area has increased more than six-fold over the last century. Today, 40% of the world's food comes from the 18% of the cropland that is irrigated. Irrigated areas increase almost 1% per year, and the irrigation water demand will increase by 13.6% by 2025. On the other hand, 8–15% of fresh water supplies will be diverted from agriculture to meet the increased demand of domestic use and industry. Furthermore, the efficiency of irrigation is very low, since only 55% of the water is used by the crop. To overcome water shortage for agriculture, it is essential to increase the water use efficiency (WUE) and to use marginal waters (reclaimed, saline, drainage) for irrigation.

Agriculture currently uses about 70% of the total water withdrawal, mainly for irrigation. In the 1980s, the global rate of increase in irrigated areas slowed considerably, mainly due to very high cost of irrigation system construction, soil salinization, the depletion of irrigation water-supplying sources, and the problems of environmental protection. Efforts are needed to find economic crops using minimal water, to use application methods that can minimize loss of water by evaporation from the soil or percolation of water beyond the depth of root zone, and to minimize losses of water from storage or delivery systems.

Under scarcity conditions, considerable efforts have been devoted over time to introduce policies aiming to increase water efficiency based on the assertion that more can be achieved with less water through better management. Better management usually refers to improvement of allocative and/or irrigation water efficiency. The former is closely related to adequate pricing, while the latter depends on the type of irrigation technology, environmental conditions, and scheduling of water application.

One of the main reasons for adopting drip irrigation in crop cultivation is to save water and increase the water use efficiency. The drip irrigation method was initially introduced in the early seventies by the agricultural universities and other research institutions in India with the aim to increase water use efficiency in crop cultivation. The development of drip irrigation was very slow in the initial years, and significant development has been achieved, especially since the 1990s. The drip irrigation system requires higher initial capital cost for installation. Because of this reason, a considerable number of farmers have expressed that they are unable to adopt this technology for low-value crops. Even though drip irrigation involves a relatively higher fixed investment, benefit-cost ratio estimation clearly suggests that the investment in drip-irrigation is economically viable to farmers, even without any subsidy by states in India. Despite availability of subsidy from state agencies, the majority of farmers are reluctant to invest in drip irrigation systems even in horticulture crops, which are highly suitable for drip irrigation. There is a need to investigate the technological options of which crop geometry modification is the most important one. Instead of adopting traditional spacing, adoption of paired row planting has been found to reduce the cost of the system by 40% in many crops including tomato, eggplant (*brinjal*), okra, etc. Therefore, drip irrigation systems should be tailored made, i.e., planned and designed based on location-specific parameters.

With objectives of analyzing the bottlenecks in the adoption of micro irrigation and to discuss various researchable issues for finding out a viable solution for the east adoption of micro irrigation, the Water Technology Centre of TNAU organized a National Congress on "New Challenges and Advances in Sustainable Micro irrigation" during March 1–3, 2017. During the technical sessions, 80 research papers were presented by scientists and research scholars, and 75 papers were presented through posters. Twenty eminent speakers delivered their keynote addresses. We thank Dr. Megh R Goyal, who was our guest speaker and took the responsibility of publishing this book volume based on selected presentations at this Congress. Following recommendations were based on the outcome of this Congress:

I. Research
- Studies to be initiated on micro irrigation with all types of automation with sensors and its suitability for various crops. An improved database on water requirements of crops under MI can be developed for future planning.
- Fertigation scheduling based on crop uptake and soil nutrient status to be developed for all the crops. Use of conventional fertilizers to be encouraged.
- Affordable micro irrigation kits may be developed for farmers with small holdings for the sustainable development of micro irrigation.
- The recycled water may be tested under drip irrigation for its feasibility.
- Studies to be taken up with cost reduction and designs for different crops.
- Studies on chemigation to be initiated.
- More physiological studies on the effect of micro irrigation on root architecture and root volume to be initiated.
- Micro irrigation recommendations to be made in a holistic way for the cropping system approach instead of recommending for individual annual short duration crop.
- Micro irrigation adoption should be linked with proper crop husbandry practices like raised bed, elite seedling, and fertigation.
- Government agencies should undertake research trials involving universities and research institutions to find out the suitability of micro irrigation for rice under groundwater irrigation in non-command

areas in a cropping system approach. A technical committee to be constituted to suggest the framework for micro irrigation for rice cultivation.

II. Extension and Dissemination of Know-How

- To make the micro irrigation sustainable in the long run, the farmers need to be empowered with capacity-building programs on layout, design, and fertigation scheduling for different crops on a continuous basis either by the research institutes or micro irrigation manufacturing firms in collaboration with public institutions.
- As the outcome of the WTC Project on drip system maintenance clearly indicated the impact of capacity building on water use efficiency and farm net returns, maintenance support is to be provided to farmers in the long run to make the water-saving technique sustainable for a considerable period.
- Under ATMA (Agriculture Technology Management Agency), the capacity-building programs may be taken up and the entire activity across the state may be coordinated by universities. In Tamil Nadu, the funding may be provided under ATMA and the project undertaken by the Water Technology Centre (WTC) of Tamil Nadu Agricultural University (TNAU) for all the drip-installed districts. Thus a state-level MI maintenance model can be developed and incorporated in the MI programs.

III. Policy and Upscaling

- Micro irrigation to be taken as a technology-driven package, not as a subsidy-driven approach. Micro irrigation implementation to be implemented by a Special Purpose Vehicle (SPV) exclusively for micro irrigation. States like Gujarat and Andhra Pradesh formed GGRC and APMIP. Similar agencies should be created in all the states for speedy adoption. All the states should develop guidelines and strict norms for the selection of micro irrigation companies and ensure the supply of quality material and proper after-sale service to farmers.
- Under PMKSY, much emphasis is being given for the promotion of micro irrigation in command areas. Pilot projects are to be commenced in all command areas before upscaling.

- Micro irrigation adoption to be focused on Mission Mode approach in selected areas where groundwater is being used for agriculture and selected crops like high-value vegetables and fruits.
- Groundwater and energy nexus can be addressed by integrating solar pump usage in agriculture.
- Financial arrangements with interest-free loans should be made for the farmers who opt for micro irrigation.
- On-farm trainings need to be conducted on different aspects on micro irrigation. Drip farmers have to be empowered with knowledge on layout, maintenance, fertigation, etc., in order to achieve the targeted benefits of micro irrigation.

The outcome of the Congress and the current book volume will help the scientists, engineers, and postgraduate students working on micro irrigation to further their knowledge on micro irrigation technology, ultimately resulting in rapid expansion of micro irrigation not only in India but throughout the world.

—B. J. Pandian, PhD
Organizing Chairman and Editor

PART I

Micro Irrigation: Principles and Challenging Technologies

CHAPTER 1

WIRELESS AUTOMATION OF A DRIP IRRIGATION SYSTEM USING CLOUD COMPUTING

M. DEIVEEGAN, R. JEYAJOTHI, S. PAZHANIVELAN, and S. RAVICHANDRAN

ABSTRACT

Water is an essential component for the development of plants in agriculture or irrigation. Excess of irrigation water not only reduces crop yield but also damages soil fertility and also causes ecological hazards like water wasting and salinity. This chapter offers a cloud-based drip irrigation technology to help the farmers; and this technology was created to enhance the watering system framework and to reduce the cost of the watering system. Sensors were placed in the farm and signals were sent continuously to wireless module frame (control station). The data was further transmitted to remote sensor hub. The proposed system aims to conserve water up to a large extent by providing automatic and manual modes of irrigation. The system can be monitored by the user on an android application.

1.1 INTRODUCTION

In the past two decades, there is immediate growth in the field of agricultural technology. Because of the highly increasing demand for freshwater, optimal usage of water resources has been provided with a greater extent by automation technology. Traditional instrumentation based on discrete and wired solutions have many difficulties to measure and control the systems, especially over the large geographical areas. Utilization of drip irrigation is reasonable and proficient in remotely monitored embedded system for irrigation purposes is essential for a farmer to economize his energy, time and money. At present, farmers have been using irrigation through labor-intensive control in which they irrigate the land at regular intervals by turning the water-pump on/off when desired. This process sometimes consumes more water and sometimes the water supply to the land is delayed due to which the crops dry-off (Singh et al., 2015).

There are many systems to achieve water savings in different crops, from basic ones to more ethnologically advanced ones. For instance, in one system plant water status was monitored and irrigation schedule was based on canopy temperature of the plant, which was acquired with thermal imaging. This project uses Arduino Uno to controls the motor and drives the pump. The Arduino Control Board is customized utilizing the Arduino IDE programming. As per the instructions are given in the program, the water pumping system is initiated (Devika et al., 2014; Guitierrez et al., 2014).

In this chapter, the programmed watering system was developed utilizing a remote sensor system (Zig-bee and internet). To enhance the watering system framework and to reduce the cost of the watering system, the technology was created. Sensors were placed in the farm and signals were sent continuously to wireless module frame (control station).

The data was further transmitted to remote sensor hub. The system was operated based on the procedure described by Feng (2011).

1.2   TECHNICAL BACKGROUND

The main theme is to control the water management for an irrigation system by the automatic method with no manual operation. The important parameters are: soil moisture sensor and the temperature sensor; the sensors sense the temperature level and soil moisture level in the atmosphere, based on transmitting the signal by ADC to the microcontroller. A microcontroller turns ON the relays to run the motor and to open the solenoid valve in a specific field and the water supply is made through the system to field.

1.3   PROPOSED HARDWARE

The proposed hardware has sensors for detecting the available soil moisture. This moisture sensor uses two probes to allow current through the soil, and then it reads that resistance to get the moisture level. More water makes the soil conduct electricity more easily (less resistance), while dry soil conducts electricity poorly (more resistance). The sensors are further connected to Aurduino via ADC. The MCU (Atmega328 built on Aurduino) is mapped with the cloud server on internet and user is given a log-in ID and password to utilize the framework with the assistance of the web. According to the message passed to the microcontroller, the Aurduino initializes the pump to water the field. If no message is passed, the system will operate automatically.

1.4   IMPLEMENTATION

The sensors are set at sufficient depth in the soil (Figures 1.1 and 1.2). The sensors are always in an active state but if the client (farmer) wishes it can be put in a passive state. Whenever sensors are in the active state they keep on communicating with the microcontroller. The microcontroller straightforwardly communicates with the driving framework (water pump) and in a roundabout way impart data to the client over the cloud stage. The MCU uses two modes for watering. One mode is the self-programmed mode, which is started if the client does not give a response within 60 minutes. Another mode is the client-driven mode, which is initiated if the user reacts to microcontroller's message. The user can control the rate of watering and can monitor his field. For the power supply of the model, solar cells were used, which were found to be cost-effective as microcontrollers devour less power. The framework was introduced in the garden of Graphic Era and was tried for four days.

1.5   FUTURE SCOPE

With the advancement of technology and cloud computing, it is easy to use a system with the help of a remote server. Therefore, integrating technology and agriculture will aid farmers to a vast degree. The era of automation

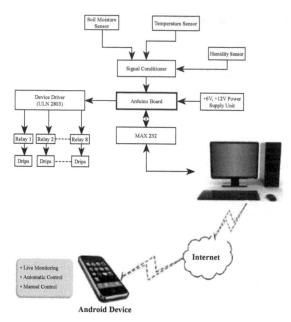

FIGURE 1.1   (See color insert.) Overview of controls and hardwares.

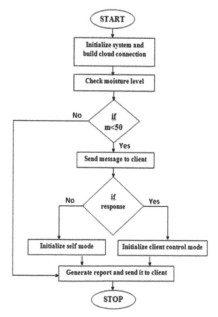

FIGURE 1.2   Flowchart for wireless automation of drip irrigation system using cloud computing.

and microcontrollers components will help the farmers to manage their resources efficiently. This will not only increase the crop production but also saves the amount of water being wasted. Moreover, implementation of microcontrollers system will be economical for the farmers as the power consumption by them is very low. In the future, if one modifies it properly, then this system can also supply agricultural chemicals like calcium, sodium, ammonium, and zinc to the field along with fertilizers by adding new sensors and valves.

1.6 SUMMARY

The developed irrigation automation system can be used in several commercial agricultural fields since it is low cost and provides a reliable operation. The application of sensor-based site-Specific irrigation has some advantages such as: preventing moisture stress of trees, diminishing of excessive water usage and rapid growing weeds. If different kinds of sensors (that is, temperature, humidity, etc.) are involved in such irrigation in future projects, it can be said that an Internet-based remote control of irrigation automation will be possible.

KEYWORDS

- **automated drip system**
- **cloud computing**
- **wireless automation**

REFERENCES

Devika, S. V., Khamuruddeen, S., Khamurunnisa, S., Thota, J., & Shaik, K., (2014). Arduino based automatic plant watering system. *Int. J. Advanced Research in Computer Science and Software Engineering, 4*(10), 449–456.

Feng, Z., (2011). Research on water-saving irrigation automatic control system based on internet of things. In: *International Conference Electric Information and Control Engineering (ICEICE)*, Wuhan – China, pp. 2541–2544.

Guitierrez, J., Medina, J. F. V., & Garibay, A. N., (2014). Automated irrigation system using a wireless sensor network and GPRS module, *IEEE Transactions on Instrumentation and Measurement, 63*(1), 166–176.

Singh, A. K., Saini, Y., & Singh, D., (2015). Cloud computing to control automatic irrigation systems. *Int. J. Advanced Research in Computer Science and Software Engineering, 5*(10), 805–807.

CHAPTER 2

FERTILIZER RECOMMENDATIONS FOR DRIP IRRIGATED COTTON UNDER AN INDUCTIVE CUM TARGETED YIELD MODEL

PRAVEENA K. STEPHEN, SANTHI RANGASAMY, and PRADIP DEY

ABSTRACT

The present investigation was conducted in transgenic cotton under drip fertigation: (i) to study the significant relationship between soil test values and crop response to fertilizers; (ii) to develop fertilizer prescription equations under IPNS (Integrated Plant Nutrition System) for desired yield target of transgenic cotton; and (iii) to test the validity of fertilizer prescription equations developed for transgenic cotton under drip fertigation. To conclude, soil test-based IPNS for desired yield targets of transgenic cotton was developed on Vertic Ustropept of Tamil Nadu under drip fertigation considering the nutrient requirements and contribution of N, P, and K from various nutrient sources (soil, fertilizer, and FYM). This envisages a balanced nutrient supply to transgenic cotton, which is site specific and can play a major component of precision agriculture.

2.1 INTRODUCTION

India has made remarkable progress in food security, poverty reduction and per capita income since the green revolution. However, the growth rate in agriculture has not kept pace with the phenomenal growth rate in industrial and service sectors. Soil fertility evaluation and fertigation are the areas, which need immediate attention since an arrest in the productivity of several crops is due to the ever-decreasing soil fertility on one hand and an imbalanced application of plant nutrients on the other. At this juncture, fertigation facilitates optimization of nutrient supply adjusted to the specific requirements of the crop at different phenological stages of growth and development.

Despite being the second largest user of fertilizers, the per hectare fertilizer use in India is still low and imbalanced. The NPK use ratio in 2009–10 was 4.3:2.0:1, which has widened to 6.7:3.1:1 in 2011–12 and it has been further distorted in 2012–13 against the desired ratio of 4:2:1 (Satish, 2013). Further, negative NPK balance in soil between crop removal and fertilizer addition has been around 8 to10 million tons per year. In addition, the recent escalation in fertilizer prices has severed a setback to the concept of balanced fertilization. The solution to these problems lies in adopting economically and ecologically sound management strategies like Soil Test Crop Response based Integrated Plant Nutrition System (STCR-IPNS) and fertigation for ensuring balanced nutrition, sustained crop productivity and soil health.

Cotton is a premier cash crop with an enormous potential of sustainable employment generation both in rural and urban sectors. Cotton lint is an important textile fiber that comprises about 35% of total world fiber use (USDA, 2011) with a cotton production area of 12.178 million ha. However, it stands second in production (35.3 million bales) (Supriya, 2010), next only to China. This shows that the potential productivity of the crop is not fully exploited. The nutrient management in cotton production is complex due to the simultaneous production of vegetative and reproductive structures during the active growth phase.

In the prevailing regime of widespread negative nutrient balances, it is difficult to expect depleted soils to support bumper crops or high yields, even in a superior hybrid or a genetically modified crop. Negative nutrient balances in most Indian soils is not only mirror of poor soil health, but also represent the severe on-going depletion of the soil's nutrient capital, degradation of the environment, and vulnerability of the crop production system in terms of its ability to sustain high yields (Tandon, 2007). Insufficient nutrient additions compared to nutrient uptake leads to a decline in soil fertility.

At this juncture, the prescription procedure outlined by Truog (1960) and modified by Ramamoorthy et al. (1967) as "*Inductive Cum Targeted Yield*

Model" provides a scientific basis for balanced fertilization and balance between applied nutrients and soil available nutrients (Ramamoorthy et al., 1967). Soil test-based fertilizer recommendation plays a vital role in ensuring balanced nutrition to crops in preventing wasteful expenditure on the use of costly mineral fertilizers. The superiority of the target yield concept over other practices for different crops gave higher yields and optimal economic returns have been reported (Khosa et al., 2012). In this context, fertilizer prescription equations for transgenic cotton were developed following the STCR-IPNS concept by adopting inductive cum targeted yield approach by Ramamoorthy et al. (1967) for Periyanaickenpalayam soil series (Vertic Ustropept) under drip fertigation.

The present investigation in this chapter was contemplated in transgenic cotton on Inceptisol under drip fertigation: (i) to elucidate the significant relationship between soil test values and crop response to fertilizers; (ii) to develop fertilizer prescription equations under IPNS for desired yield target of transgenic cotton; and (iii) to test verify the validity of fertilizer prescription equations developed for transgenic cotton under drip fertigation.

2.2   MATERIALS AND METHODS

Studies on Soil Test Crop Response based Integrated Plant Nutrition System were conducted adopting the Inductive cum Targeted Yield Model, on a Vertic Ustropept of Tamil Nadu, India.

The methodology adopted in this study is based on the procedure outlined by Truog (1960) and modified by Ramamoorthy et al. (1967) as "Inductive cum Targeted yield model" which provides a scientific basis for balanced fertilization and balance between applied nutrients and soil available nutrients forms. The operational range of variation in soil fertility was created deliberately to generate data covering an appropriate range of values for each controllable variable (fertilizer dose) at different levels of an uncontrollable variable (soil fertility), which could not be expected to occur at one place normally. Hence to create fertility variations in the same field, a gradient experiment was conducted prior to the test crop experiment to reduce the heterogeneity in the soil population studied, management practices adopted and climatic conditions prevailing.

2.2.1   STUDY SITE AND SOIL DESCRIPTION

The field experiments were conducted at the Eastern block of Tamil Nadu Agricultural University Farm, Coimbatore, Tamil Nadu on Inceptisol (Vertic Ustropept). The farm is located in Western agro-climatic zone of Tamil Nadu at 11°12" North latitude and 77° 03" East longitude at an altitude of 426.74 m above MSL. The gradient and test crop experiments were conducted during October 2011 to April 2012. The soil at the experimental field belongs to Periyanaickenpalayam series taxonomically referred to as Vertic Ustropept exhibiting clay loam texture, moderately alkaline reaction (pH 8.4) and non – saline conditions (EC 0.17 dS m^{-1}). The initial soil fertility status showed low organic carbon (4.7 g kg^{-1}), low available N (225 kg ha^{-1}), medium available P (19.9 kg ha^{-1}), and high available K (570 kg ha^{-1}). The available Zn, Cu, and Mn were in the sufficient range (1.29, 1.94 and 11.39 mg kg^{-1}, respectively), while available Fe was in the deficient range (3.34 mg kg^{-1}). The total N, P and K contents in the soil were 0.13, 0.09 and 0.45%, respectively. The P and K fixing capacities of the soil were 100 kg ha^{-1}.

2.2.2   TREATMENT STRUCTURE AND SOIL AND PLANT ANALYSIS

The field experiments consisted of fertility gradient experiment with fodder maize (var. CO–1) and the test crop experiment with transgenic cotton (RCH–530-BGII); and were conducted at TNAU Farm, Coimbatore on Inceptisol. The approved treatment structure and layout design (Figure 2.1) as followed in the All India Coordinated Research Project for Investigations on Soil Test Crop Response Correlation (AICRP-STCR) based on "Inductive cum Targeted yield model" (Ramamoorthy et al., 1967) was adopted in the present investigation.

2.2.2.1   GRADIENT EXPERIMENT

In the gradient experiment, an operational range of variation in soil fertility was created deliberately. For this purpose, the experimental field was divided into three equal strips, the first strip received no fertilizer ($N_0P_0K_0$), the second and third strips received one ($N_1P_1K_1$) and two ($N_2P_2K_2$) times the standard dose of N, P_2O_5, and K_2O, respectively and a gradient crop of fodder maize (*var.* CO–1) was grown. Eight pre-sowing and post-harvest soil samples were collected from each fertility strip and analyzed for alkaline $KMnO_4$-N (Subbiah, 1956), Olsen–P (Olsen et al., 1954) and NH_4OAc-K (Stanford et al., 1949). At harvest, plant samples were collected, processed and analyzed for N (Humphery, 1956), P and K contents (Jackson, 1973); and NPK uptake was computed.

2.2.2.2 TEST CROP EXPERIMENT

After confirming the establishment of fertility gradients in the experimental field, for the second phase of the field experiment, each strip was divided into 24 plots (Figure 2.1), and initial soil samples were collected from each plot and analyzed for alkaline KMnO4-N, Olsen-P, and NH4OAc-K. The experiment was laid out in a fractional factorial design comprising 24 treatments and the test crop experiment with cotton was conducted with four levels each of N (0, 60, 120 and 180 kg ha^{-1}), P_2O_5 (0, 30, 60 and 90 kg ha^{-1}) and K_2O (0, 40, 80 and 120 kg ha^{-1}) and three levels of FYM (0, 6.25 and 12.5 t ha^{-1}). The experiment was conducted as per the approved guidelines of AICRP-STCR and fertilizer recommendations were developed.

The IPNS treatments were NPK alone, NPK+ FYM @ 6.25 t ha^{-1} and NPK+ FYM @ 12.5 t ha^{-1}, and were superimposed across the strips. There were 21 fertilizer treatments along with three controls, which were randomized in each strip in such a way that all the treatments occurred in both the directions. The treatment structure and layout are given in Figure 2.1. Routine cultural operations were followed periodically. Fertilizer doses were imposed as per the treatments and the fertigation was done at weekly intervals as per the schedule finalized by for cotton on Inceptisol (Jayaprakash, 2008). The sources of nutrients used in fertigation were urea, single superphosphate, and Muriate of Potash. The seed cotton, plant, and post-harvest soil samples were collected from each plot and analyzed; and NPK uptake by cotton was computed using the dry matter yield.

	STRIP I		STRIP II		STRIP III		
	$N_0P_0K_0$	$N_2P_3K_2$	$N_0P_0K_0$	$N_3P_2K_3$	$N_0P_0K_0$	$N_2P_2K_1$	
	$N_2P_2K_0$	$N_2P_1K_1$	$N_3P_2K_1$	$N_2P_2K_3$	$N_3P_3K_1$	$N_1P_1K_1$	NPK alone
	$N_3P_1K_3$ (A)	$N_1P_1K_2$	$N_2P_0K_2$ (B)	$N_1P_2K_1$	$N_0P_2K_2$ (C)	$N_2P_3K_3$	B I
	$N_1P_2K_2$	$N_3P_3K_2$	$N_2P_2K_2$	$N_3P_3K_3$	$N_3P_2K_2$	$N_2P_1K_2$	
	$N_0P_0K_0$	$N_3P_2K_3$	$N_0P_0K_0$	$N_2P_2K_1$	$N_0P_0K_0$	$N_2P_3K_2$	NPK+
	$N_3P_2K_1$	$N_2P_2K_3$	$N_3P_3K_1$	$N_1P_1K_1$	$N_2P_2K_0$	$N_2P_1K_1$	6.25 t ha^{-1} FYM
OUTS	$N_2P_0K_2$ (B)	$N_1P_2K_1$	$N_0P_2K_2$ (C)	$N_2P_3K_3$	$N_3P_1K_3$ (A)	$N_1P_1K_2$	B II
	$N_2P_2K_2$	$N_3P_3K_3$	$N_3P_2K_2$	$N_2P_1K_2$	$N_1P_2K_2$	$N_3P_3K_2$	
	$N_0P_0K_0$	$N_2P_2K_1$	$N_0P_0K_0$	$N_2P_3K_2$	$N_0P_0K_0$	$N_3P_2K_3$	NPK+
	$N_3P_3K_1$	$N_1P_1K_1$	$N_2P_2K_0$	$N_2P_1K_1$	$N_3P_2K_1$	$N_2P_2K_3$	12.5 t ha^{-1} FYM
	$N_0P_2K_2$ (C)	$N_2P_3K_3$	$N_3P_1K_3$ (A)	$N_1P_1K_2$	$N_2P_0K_2$ (B)	$N_1P_2K_1$	B III
	$N_3P_2K_2$	$N_2P_1K_2$	$N_1P_2K_2$	$N_3P_3K_2$	$N_2P_2K_2$	$N_3P_3K_3$	

Treatment structure

1. $N_0P_0K_0$	5. $N_1P_1K_1$	9. $N_2P_1K_1$	18. $N_3P_1K_1$
2. $N_0P_0K_0$	6. $N_1P_1K_2$	10. $N_2P_0K_2$	19. $N_3P_2K_1$
3. $N_0P_0K_0$	7. $N_1P_2K_1$	11. $N_2P_1K_2$	20. $N_3P_2K_2$
4. $N_0P_1K_2$	8. $N_1P_2K_3$	12. $N_2P_2K_0$	21. $N_3P_3K_1$
		13. $N_2P_2K_1$	22. $N_3P_3K_2$
		14. $N_2P_2K_2$	23. $N_3P_2K_3$
		15. $N_2P_3K_1$	24. $N_3P_3K_3$
		16. $N_2P_3K_2$	
		17. $N_2P_3K_3$	

FIGURE 2.1   (See color insert.) Layout plan of STCR –IPNS experiment with transgenic cotton under drip fertigation (Field 76, Eastern block, TNAU, CBE).

2.2.3   BASIC PARAMETERS FOR FERTILIZER PRESCRIPTION EQUATIONS

Making use of data on the yield of cotton, total uptake of N, P and K, initial soil test values for available N, P and K and doses of fertilizer N, P_2O_5 and K_2O applied, the basic parameters *viz.*, nutrient requirement (NR), contribution of nutrients from soil (Cs), fertilizer (Cf) and farmyard manure (Cfym) were calculated as outlined by Ramamoorthy et al. (1967).

2.2.3.1   NUTRIENT REQUIREMENT

The kg of N/P_2O_5/K_2O required per quintal (100 kg) of seed cotton production, expressed in kg (100 kg)$^{-1}$ was calculated as follows:

$$NR = [(\text{Total uptake of N or } P_2O_5 \text{ or } K_2O \text{ (kg ha}^{-1}))]/$$
$$[\text{Seed cotton yield } ((100 \text{ kg) ha}^{-1})] \tag{1}$$

2.2.3.2   PERCENT CONTRIBUTION OF NUTRIENTS FROM SOIL TO TOTAL NUTRIENT UPTAKE (CS)

$$Cs = 100 \text{ x } [(\text{Total uptake of N or } P_2O_5 \text{ or } K_2O$$
$$\text{in control plot (kg ha}^{-1}))]/$$
$$[(\text{Soil test value for available N or } P_2O_5 \text{ or } K_2O$$
$$\text{in control plot (kg ha}^{-1}))] \tag{2}$$

2.2.3.3   PERCENT CONTRIBUTION OF NUTRIENTS FROM FERTILIZER TO TOTAL UPTAKE (CF)

$$Cf = 100 \text{ x } \{[(\text{Total uptake of N or } P_2O_5 \text{ or } K_2O$$
$$\text{in treated plot (kg ha}^{-1})) - (\text{Soil test value for available}$$
$$\text{N or } P_2O_5 \text{ or } K_2O \text{ in control plot (kg ha}^{-1}) * \text{Average Cs})]$$
$$/[\text{Fertilizer N or } P_2O_5 \text{ or } K_2O \text{ applied (kg ha}^{-1})]\} \tag{3}$$

2.2.3.4   PERCENT CONTRIBUTION OF NUTRIENTS FROM ORGANICS TO TOTAL UPTAKE (CO)

2.2.3.4.1   Percent Contribution from FYM (Cfym)

$$Cfym = 100 \text{ x } \{[(\text{Total uptake of N or P or K in}$$
$$\text{FYM treated plot (kg ha}^{-1})) - (\text{Soil test value for available}$$
$$\text{N or P or K in FYM treated plot (kg ha}^{-1}) * \text{Average Cs})]/$$
$$[\text{Nutrient N/P/K added through FYM (kg ha}^{-1})]\} \tag{4}$$

These parameters were used for developing fertilizer prescription equations for deriving fertilizers doses; and the soil test-based fertilizer recommendations were prescribed in the form of a ready table for desired yield target of cotton under NPK alone and under IPNS.

2.2.4 FERTILIZER PRESCRIPTION EQUATIONS

Fertilizer nitrogen (FN):

$$FN = \{[(NR/(Cf/100))*T] - [(Cs/Cf)*SN]\} \tag{5}$$

$$FN = \{[(NR/(Cf/100))*T] - [(Cs/Cf)*SN] - [(Cfym/Cf)*ON]\} \tag{6}$$

Fertilizer phosphorus (FP$_2$O$_5$):

$$FP2O5 = \{[(NR/(Cf/100))*T] - [(Cs/Cf)*2.29SP]\} \tag{7}$$

$$FP2O5 = \{[(NR/(Cf/100))*T] - [(Cs/Cf)*2.29SP]$$
$$- [(Cfym/Cf)*2.29OP]\} \tag{8}$$

Fertilizer potassium (FK$_2$O):

$$FK2O = \{[(NR/(Cf/100))*T] - [(Cs/Cf)*1.21SK]\} \tag{9}$$

$$FP2O5 = \{[(NR/(Cf/100))*T] - [(Cs/Cf)*2.29SK]$$
$$- [(Cfym/Cf)*1.21OK]\} \tag{10}$$

Where, FN, FP$_2$O$_5$, and FK$_2$O are fertilizer N, P$_2$O$_5$ and K$_2$O in kg ha^{-1}, respectively; NR is a nutrient requirement (N or P$_2$O$_5$ or and K$_2$O) in kg (100 kg)$^{-1}$; Cs is percent contribution of nutrients from soil; Cf is percent contribution of nutrients from fertilizer; Cfym is percent contribution of nutrients from FYM; T is the yield target in (100 kg) ha^{-1}; SN, SP, and SK, respectively, are alkaline KMnO$_4$-N, Olsen-P, and NH$_4$OAc-K in kg ha^{-1}; and ON, OP, and OK are the quantities of N, P, and K supplied through FYM in kg ha^{-1}.

These equations serve as basis for predicting fertilizer doses for specific yield targets (T) of cotton for varied soil available nutrient levels.

2.3 RESULTS AND DISCUSSION

2.3.1 SEED COTTON YIELD, UPTAKE AND INITIAL AVAILABLE NPK STATUS

The range and mean values indicated that the seed cotton yield ranged from 1082 kg ha^{-1} in absolute control to 3405 kg ha^{-1} in $N_{180}P_{90}K_{80}$ + FYM @ 12.5 t ha^{-1} of strip II with mean values of 2146, 2691 and 2803 kg ha^{-1}, respectively in strips I, II and III (Table 2.1). The N uptake by cotton varied from 43.2 to 152.9 kg ha^{-1}; P uptake from 8.69 to 47.7 kg ha^{-1} and K uptake from 52.2 to 140.2 kg ha^{-1} in strips I, II and III, respectively.

The data on initial soil test values of cotton revealed that the mean KMnO$_4$-N was 213, 238 and 255 kg ha^{-1}, respectively in strips I, II and III. The mean Olsen-P values were 16.4, 30.4 and 38.0 kg ha^{-1}, respectively in strips I to III and the mean NH$_4$OAc-K values were 554, 589 and 609 kg ha^{-1} in strips I, II and III, respectively (Table 2.1).

The existence of operational range of soil test values for available N, P and K status in the present investigation was clearly depicted from the initial soil available nutrient status and variations in the seed cotton yield of cotton and NPK uptake, which is a prerequisite for calculating the basic parameters and developing fertilizer prescription equations for calibrating the fertilizer doses for specific yield target of cotton. The similar existence of an operational range of available N, P, and K for sunflower on Inceptisol was reported (Andi, 1998).

TABLE 2.1 Initial Soil Available NPK, Fruit Yield, and NPK Uptake by Cotton (kg ha^{-1})

Parameters (kg ha^{-1})	Strip-I		Strip-II		Strip-III	
	Range	Mean	Range	Mean	Range	Mean
KMnO$_4$ –N	207–216	213	232–241	238	252–260	255
Olsen–P	15–18	16.4	28–33	30.4	36–42	38.0
NH$_4$OAc-K	550–560	554	584–594	589	606–613	609
Seed cotton yield	1082–2618	2146	1275–3405	2691	1406–3401	2803
N uptake	43.2–117.1	94.0	57.8–152.9	118.0	63.83–152.6	124.5
P uptake	8.69–24.9	19.5	13.1–47.7	27.6	13.0–46.6	28.4
K uptake	52.2–103.4	84.2	62.2–140.2	109.9	69.9–139.7	114.8

2.3.2   RESPONSE OF TRANSGENIC COTTON TO FERTILIZER N, P_2O_5, AND K_2O

The response of cotton to different levels of fertilizer N, P_2O_5 and K_2O were assessed in terms of response ratio (RR). There was a progressive increase in response for N up to the highest level *i.e.* 180 kg ha^{-1} and the highest RR recorded was 5.30 at N_{180}. A similar trend was observed for phosphorus and potassium with the highest RR of 4.60 and 2.95 observed at P_{90} and K_{120} respectively (Table 2.2). Application of N, P and K had a significant effect on plant growth and yield and there was a progressive increase in response for N, P_2O_5 and K_2O levels from N_{60} to $N_{180,}$ P_{30} to P_{90} and K_{40} to K_{120}, respectively.

2.3.3   BASIC PARAMETERS

In the targeted yield model, the basic parameters for developing fertilizer prescription equations for cotton are: (i) nutrient requirement (NR) in kg per 100 kg of seed cotton; (ii) percent contribution of available NPK from soil (Cs), fertilizers (Cf) and farmyard manure (Cfym). Making use of data on the yield of cotton, total uptake of N, P and K, initial soil test values for available N, P and K and doses of fertilizer N, P_2O_5 and K_2O applied, the basic parameters were computed.

TABLE 2.2   Response of Transgenic Cotton to Different Levels of Fertilizer Nutrients

Nitrogen (N)			Phosphorus (P_2O_5)			Potassium (K_2O)		
Level (kg ha^{-1})	Response (kg)	Response Ratio (kg kg^{-1})	Level (kg ha^{-1})	Response (Kg)	Response Ratio (kg kg^{-1})	Level (kg ha^{-1})	Response (kg)	Response Ratio (kg kg^{-1})
60	252	4.20	30	92	3.07	40	97	2.43
120	572	4.77	60	271	4.52	80	214	2.68
180	956	5.30	90	413	4.60	120	354	2.95

Application of adequate amount of nutrients is a pre-requisite for exploiting genetic potential of any crop. Cotton which is a heavy feeder exhibits vigorous growth and dry matter production (DMP) and is responsive to application of fertilizers. The nutrient requirement to produce one quintal (100 kg) of seed cotton was 4.43 kg of N, 2.20 kg of P_2O_5 and 4.83 kg of K_2O (Table 2.3). In the present investigation, the requirement of K_2O was higher followed by N and P_2O_5. The requirement of K_2O was 1.09 times higher than N and 2.20 times higher than P_2O_5. A similar trend of the nutrient requirement for N, P_2O_5, and K_2O was also reported for rainfed transgenic cotton on black calcareous soil (Anonymous, 2011), for rainfed cotton (var. Narasimha) on a Vertisol (Subha-Rao et al., 1956). The affinity of cotton towards potassium has also been reported by Jagvir et al. (2000).

TABLE 2.3   Nutrient Requirement, Percent Contribution of Nutrients from Soil, Fertilizer, and FYM for Transgenic Cotton

Parameters	Basic Data		
	N	P_2O_5	K_2O
Nutrient requirement (kg q^{-1})	4.43	2.20	4.83
Percent contribution from soil (Cs)	24.65	48.95	11.06
Percent contribution from fertilizers (Cf)	52.01	49.89	73.35
Percent contribution from FYM (Cfym)	38.19	16.43	40.35

The percent contribution of nutrients from the soil (Cs) to the total uptake was computed from the absolute control plots and it expresses the capacity of the crop to extract nutrients from the soil. In the present study, it was found that the soil had contributed 24.65% of available N, 48.95% of available P and 11.06% of available K respectively towards the total N, P and K uptake by cotton (Coumaravel, 2012). The nutrient contribution of the soil to transgenic cotton was the highest for P compared to that by N and K. Regarding N and K, comparatively lower Cs was recorded due to the preferential nature of cotton towards the applied N and K_2O than the native N and K. This is in accordance with the study in Maharashtra on transgenic cotton var. Mallika (Muralidharudu et al., 2007) and for rainfed cotton on a Vertisol (Subha-Rao et al., 1956).

The percent contribution from fertilizer nutrients (Cf) towards the total uptake by cotton was 52.01, 49.89 and 73.35%, respectively for N, P_2O_5 and K_2O (Table 2.3) and followed the order of $K_2O > N > P_2O_5$. The estimated percent contribution of nutrients from fertilizers (Cf) to total uptake clearly revealed the fact that the magnitude of contribution by fertilizer

K_2O was 1.47 times higher than P_2O_5 and 1.41 times as that of N. The contribution from fertilizers was higher than from the soil for all the three nutrients. A similar trend for transgenic cotton hybrid BRAHMA on black calcareous soil has been reported (Anonymous, 2011). The contribution of nutrients towards the growth of the crop was higher from fertilizers than that of soil for all the three nutrients (N, P_2O_5, and K_2O). A similar trend of results for jute in West Bengal and wheat on an Inceptisol in Punjab has been observed (Muralidharudu et al., 2007).

The estimated percent contribution of N, P_2O_5 and K_2O from FYM (Cfym) was 38.19, 16.43 and 40.35%, respectively for cotton, which indicated that relatively higher contribution was recorded for K_2O followed by N and P_2O_5 for cotton. The response yardstick recorded was 5.13 kg kg^{-1}. Similarly, the contribution of nutrients from FYM for cotton also indicated that relatively higher contribution was recorded for K_2O followed by N and P_2O_5. These findings corroborated with the earlier findings on Ashwagandha (Saranya et al., 2012) and onion (Santhi et al., 2002).

2.3.4   FERTILIZER PRESCRIPTION EQUATIONS FOR TRANSGENIC COTTON

Soil test-based fertilizer prescription equations for desired yield target of cotton were formulated using the basic parameters and are given below:

STCR-NPK alone

$$FN = 8.51\ T - 0.47\ SN$$
$$FP_2O_5 = 4.41\ T - 2.25\ SP$$
$$FK_2O = 6.59\ T - 0.18\ SK \tag{11}$$

STCR-IPNS (NPK + FYM)

$$FN = 8.51\ T - 0.47\ SN - 0.73\ ON$$
$$FP_2O_5 = 4.41T - 2.25\ SP - 0.75\ OP$$
$$FK_2O = 6.59\ T - 0.18\ SK - 0.66\ OK \tag{12}$$

where: FN, FP_2O_5, and FK_2O are fertilizer N, P_2O_5 and K_2O in kg ha^{-1}, respectively; T is the yield target in quintal ha^{-1} (1 quintal = 100 kg); SN, SP and SK respectively are alkaline $KMnO_4$-N, Olsen-P, and NH_4OAc-K in kg ha^{-1}; and ON, OP and OK are quantities of N, P and K supplied through FYM in kg ha^{-1}.

Fertilizer response is denoted by the functional relationship between increase in crop yield and added fertilizers. It can be expressed graphically by a curve or algebraically by an equation. The superiority of the target yield concept over other practices for different crops as it gave higher yields, net benefit and optimal economic returns has been reported (Milap et al., 2006). The yield targets were achieved within reasonable limits when the fertilizer was applied based on soil test in the majority of the crops thus establishing the utility of the prescription equations for recommending soil test-based fertilizer application to the farmers. With this background in the present investigation, Soil test-based fertilizer prescription equations for the desired yield target of cotton were developed using the basic parameters obtained. The data clearly revealed that the fertilizer N, P_2O_5, and K_2O requirements were decreased with increase in soil test values and were increased with the increase in yield targets.

Realizing the superiority of the targeted yield approach, documentation has been done in a handbook on the soil test and yield target based integrated fertilizer prescriptions, for a range of 44 soil-crop situations in Tamil Nadu, which includes cereals, millets, oilseeds, sugarcane, cotton, vegetables, spices and medicinal plants (Santhi et al., 2012).

2.3.5   FERTILIZER PRESCRIPTION UNDER IPNS FOR DESIRED YIELD TARGET OF TRANSGENIC COTTON

A ready reckoner table was prepared using equations in Section 2.3.4 for a range of soil test values and for yield targets of 3.0 and 3.5 t ha^{-1} of seed cotton (Table 2.4). For achieving a yield target of 3.0 t ha^{-1} of seed cotton with soil test values of 280, 20, and 500 kg ha^{-1} of $KMnO_4$-N, Olsen-P and NH_4OAc-K, the fertilizer N, P_2O_5, and K_2O doses required were 124, 87, and 108 kg-ha^{-1}, respectively under NPK alone and 84, 67, and 74 kg ha^{-1} under IPNS (NPK + FYM @ 12.5 t ha^{-1} with 32, 0.64, 0.31, and 0.61% of moisture, N, P, and K, respectively).

TABLE 2.4 Soil Test Based Fertilizer Prescription for Yield Targets of 3.0 and 3.5 t ha^{-1} of Transgenic Cotton (kg ha^{-1})

Soil test values	Treatments					
	NPK alone (kg ha^{-1})	NPK+ FYM 12.5 t ha^{-1} (kg ha^{-1})	Percent reduction over NPK	NPK alone (kg ha^{-1})	NPK+ FYM 12.5 t ha^{-1} (kg ha^{-1})	Reduction over NPK (%)
Yield target = 3000 kg/ha or 3.0 t ha^{-1}				Yield target = 3500 kg/ha or 3.5 t ha^{-1}		
KMnO$_4$-N (kg ha^{-1})						
160	180	140	22.2	223	183	18.0
180	171	131	23.4	213	173	18.8
200	161	121	24.8	204	164	19.6
220	152	112	26.3	194	154	20.6
240	143	103	28.1	185	145	21.6
260	133	93	30.1	176	136	22.8
280	124	84	32.3	166	126	24.1
Olsen-P (kg ha^{-1})						
10	110	90	18.2	132	112	15.2
12	105	85	19.0	127	107	15.7
14	101	81	19.8	123	103	16.3
16	96	76	20.8	118	98	16.9
18	92	72	21.8	114	94	17.6
20	87	67	22.9	109	89	18.3
22	83	63	24.2	105	85	19.1
NH$_4$OAC-K (kg ha^{-1})						
300	144	110	23.7	177	143	19.2
350	135	101	25.2	168	134	20.3
400	126	92	27.0	159	125	21.4
450	117	83	29.1	150	116	22.7
500	108	74	31.6	141	107	24.2
550	99	65	34.4	132	98	25.8
600	90	56	37.9	123	89	27.7

Similarly, for the target of 3500 kg ha^{-1}, the respective values were 166, 109, and 141 kg ha^{-1} under NPK alone and 126, 89, and 107 under IPNS. Under IPNS, the fertilizer savings were 40, 20, and 34 kg ha^{-1}, respectively when FYM was applied @12.5 t ha^{-1} along with NPK fertilizers.

In the present investigation, there was a marked response to the application of NPK fertilizers, and the magnitude of response was higher under IPNS compared to NPK alone. The percent reduction in NPK fertilizers under IPNS was also increased with increasing soil fertility levels with reference to NPK and was decreased with the increase in yield targets. These could be achieved by integrated use of FYM with NPK fertilizers. The role of FYM is multidimensional, ranging from building up of organic matter, maintaining favorable soil physical properties, and a balanced supply of nutrients. In the present investigation also, these factors might have contributed to the yield enhancement in cotton, when NPK fertilizers are coupled with FYM. A similar trend of results was reported in maize (Coumaravel, 2012) and in transgenic cotton (Anonymous, 2011).

2.4 SUMMARY

Studies on Soil Test Crop Response based Integrated Plant Nutrition System (STCR – IPNS) were conducted during 2011–12 adopting the Inductive cum Targeted yield model, on a Vertic Ustropept soil of Tamil Nadu to develop fertilizer prescriptions for the desired yield targets of transgenic cotton through drip fertigation. The basis for making the fertilizer prescriptions viz. nutrient requirement (NR), the contribution of nutrients from the soil (Cs), fertilizer (Cf) and farmyard manure (Cfym) were computed using the field experimental data. Making use of these basic parameters, the fertilizer prescription equations were developed under NPK alone and IPNS. Using the equations, nomograms were formulated for a range of soil test values under NPK alone and under IPNS for desired yield target of cotton. When NPK was applied along with FYM @ 12.5 t ha^{-1}, the extent of saving was 40, 20, and 34 kg of fertilizer N, P_2O_5, and K_2O respectively for cotton resulting in economy of fertilizer use under IPNS.

KEYWORDS

- drip fertigation
- farmyard manure
- inceptisol
- IPNS
- seed cotton
- STCR
- targeted yield model
- transgenic cotton

REFERENCES

Andi, K., (1998). Soil test crop response studies under integrated plant nutrition system for Okra – Sunflower cropping sequence on Inceptisol. *PhD Dissertation Submitted to Tamil Nadu Agricultural University (TNAU),* Coimbatore, p. 203.

Anonymous, (2011). Progress report of the All India Coordinated Research Project for Investigation on Soil Test Crop Response Correlation. Tamil Nadu Agricultural University, Coimbatore, pp. 28–44.

Coumaravel, K., (2012). Soil test crop response correlation studies through integrated plant nutrition system for maize – tomato sequence. *PhD (Agronomy) Dissertation Submitted Tamil Nadu Agricultural University (TNAU)*, Coimbatore, p. 185.

Humphry, E. C., (1956). Mineral components and ash analysis. In: *Modern Methods of Plant Analysis*, Springer – Verlag, Berlin, *1*, 468–562.

Jackson, M. L., (1973). *Soil Chemical Analysis*. Prentice Hall of India Private Ltd., New Delhi, p. 498.

Jagvir, S., & Blaise, D., (2000). *Nutrient Management in Rainfed Cotton*. CICR technical bulletin 6, Central Institute for Cotton Research, Nagpur, p. 28.

Jayaprakash, N., (2008). *Nutrient Dynamics Under Drip Fertigation in Cotton*. M. Sc. (Agronomy) thesis submitted to Tamil Nadu Agricultural University, Coimbatore, p. 125.

Khosa, M. K., Sekhon, B. S., Ravi, M. S., Benipal, D. S., & Benbi, D. K., (2012). Performance of target yield based fertilizer prescription equations in rice-wheat cropping system in Punjab. *Indian J. Fert.*, *8*(2), 14–18.

Milap, C., Benbi, D. K., & Benipal, D. S., (2006). Fertilizer recommendations based on soil tests for yield targets of mustard and rapeseed and their validations under farmer's field conditions in Punjab. *J. Indian. Soc. Soil Sci.*, *54*(3), 316–321.

Muralidharudu, Y., Rathore, A., & Subba-Rao, A., (2007). *18th Progress report of All India Coordinated Project for Investigation on Soil Test Crop Response Correlation*. Indian Institute of Soil Science, Bhopal, p. 215.

Olsen, S. R., Cole, C. V., Watanabe, F. S., & Dean, L., (1954). *Estimation of Available Phosphorus in Soils by Extraction with Sodium Bicarbonate*. United States Department of Agriculture Circular 939, U. S. Govt. Printing Office, Washington, DC, p. 21.

Ramamoorthy, B., Narasimham, R. K., & Dinesh, R. S., (1967). Fertilizer application for specific yield targets on Sonora 64 (wheat). *Indian Fmg., 17*, 43–45.

Santhi, R, Natesan, R., & Selvakumari, G., (2002). Soil test crop response correlation studies under integrated plant nutrition system for onion (*Allium cepa L. var. aggregatum*) in Inceptisol of Tamil Nadu. *J. Indian Soc. Sci., 50*, 489–492.

Santhi, R., Maragatham, S., Sellamuthu, K. M., Natesan, R., Duraisami, V. P., Dey, P., & Rao, S. A., (2012). *Handbook on Soil Test and Yield Target-based Integrated Fertilizer Prescriptions*. AICRP-STCR, Tamil Nadu Agricultural University, Coimbatore, p. 221.

Saranya, S., Santhi, R., Appavu, K., & Rajamani, K., (2012). Soil test-based integrated plant nutrition system for Ashwagandha on Inceptisols. *Indian J Agric Res., 46*, 88–90.

Satish, C., (2013). Adoption of fertilizer best management practice. *Indian J. Fert., 9*(4), 10–11.

Stanford, S., & English, L., (1949). Use of flame photometer in rapid soil tests of K and Ca. *Agron J., 41*, 446–450.

Subba-Rao, A., & Rathore, A., (2003). *17th Progress Report of the All India Coordinated Research Project for Investigation on Soil Test Crop Response Correlation*. Indian Institute of Soil Science, Bhopal, pp. 14, 15.

Subbiah, B. V., & Asija, G. L., (1956). A rapid procedure for the estimation of available nitrogen in soils. *Curr Sci., 25*, 259–260.

Supriya, P., (2010). Indian cotton production: Current scenario. *The Indian Textile J.*, http://indiantextilejournal.com/articles/FAdetails.asp?id=2737 Accessed July 31, 2018.

Tandon, H. L. S., (2007). Soil nutrient balance sheets in India: Importance, status, issues, and concerns. *Better Crops - India*, 15–19.

Truog, E., (1960). Fifty years of soil testing. Trans 7th Intl. *Congress Soil Sci. Vol. III Commission IV Paper 7*, Almaty – Kazakhatan, pp. 46–53.

USDA (United States Department of Agriculture), (2011). Economic Research Service, http://www.ers.usda.gov/Briefing/Cotton/ Accessed July 31, 2018.

CHAPTER 3

SUPPLEMENTARY IRRIGATION USING RAIN-GUN SPRINKLERS FOR INCREASING PRODUCTIVITY IN DRYLANDS

S. SOMASUNDARAM

ABSTRACT

This chapter focuses on suitable minimum tillage + crop residue and supplementary irrigation options for increasing dryland productivity in Tamil Nadu. Therefore, the field experiments were conducted at Agricultural Engineering College and Research Institute, Kumulur, Tamil Nadu, India during 2013–2015. Supplementary irrigation was given using rain-gun sprinklers from harvested rainwater in the farm pond. The test crop was black gram var. VBN-(BG)–6. The harvested rainwater in the farm pond was 1350 m^3 and 920 m^3 with a rainfall of 336 mm and 320 mm during 2013–14 and 2014–15, respectively. The results revealed that water used in control without irrigation, one, two, and three supplementary irrigation treatments was 269, 299, 319 and 339 mm during 2013–14; and 256, 276, 296, and 316 mm during 2014–15, respectively. Minimum tillage + crop residue @ 5 t ha^{-1} with three supplementary irrigations through rain-gun sprinklers registered higher black gram grain yield (986 kg ha^{-1} during 2013–14 and 911 kg ha^{-1} during 2014–15) and was comparable with the combined effect of minimum tillage + crop residue @ 5 t ha^{-1} with two supplementary irrigation through rain-gun sprinklers during both years of study. Therefore, this recommendation is extended for sustaining productivity in drylands of Tamil Nadu.

3.1   INTRODUCTION

Increase in dryland productivity may be achieved through identifying technologies, which facilitate water conservation and efficient use of limited rainwater in drylands. Practicing conservation agriculture (combining minimum tillage and crop residue) and use of harvested rainwater for supplementary irrigation through micro irrigation systems may be superior options to mitigate climate change and increase yield in drylands (Delgado, 2010; Oweis, 1997). Harvesting rainwater in farm ponds and efficient use for supplementary irrigation for increasing yield has been investigated (Krishna et al., 1987; Wang, 2017). Now rain-gun sprinklers are gaining momentum among farmers and success of the system in providing supplementary irrigation in drylands may increase crop yield (Mostafa et al., 2011; Somasundaram et al., 2011).

However, limited attempts have been made in identifying the best combination of these technologies in dryland tracts of Tamil Nadu. Therefore, this chapter focuses on suitable minimum tillage + crop residue and supplementary irrigation options for increasing dryland productivity in Tamil Nadu.

3.2   MATERIALS AND METHODS

Two field experiments were conducted at Research Farm of Agricultural Engineering College and Research Institute, Kumulur, Tamil Nadu, India during North-East monsoon season of 2013–2015 under dryland situations. Blackgram variety VBN (BG) 6 was used in the study. The total quantity of 336 mm and 320 mm of rainfall was received in 21 and 25 rainy days during 2013–14 and 2014–15 cropping season, respectively. Effective rainfall was determined by the water balance method suggested by Dastane (2010).

The soil at the experimental field was sandy loam with pH of 7.8, organic carbon 0.53%, available nitrogen (186 kg ha^{-1}), available phosphorus (18 kg ha^{-1}), available potassium (215 kg ha^{-1}), EC of 0.56 dSm^{-1}, WHC of 57.1% and IR 13.2 cm h^{-1}. Rainwater was harvested in the pond with a capacity 1350 m^3. The harvested water was used for supplementary irrigation using rain-gun sprinklers. The rain-gun sprinkler system with a discharge rate of 8–10 lps at operating pressure of 8 bars and wetting diameter of 20–25m was designed as per specific treatment. The rain-gun was attached to 5 HP diesel engine from the farm pond with harvested rainwater for providing supplementary irrigation. The experiment was laid out in a split plot design with four main plots and four subplots as follows:

M_1 – conventional tillage;
M_2 – minimum tillage without crop residue;
M_3 – minimum tillage + crop residue @ 2.5 t ha^{-1}; and
M_4 – minimum tillage + crop residue @ 5 t ha^{-1}.
S_1 – one supplementary irrigation through rain-gun sprinklers;
S_2 – two supplementary irrigation through rain-gun sprinklers;
S_3 – three supplementary irrigation through rain-gun sprinklers; and
S_4 – no supplementary irrigation.

The treatments were replicated thrice. The data on supplementary irrigation, water used and yield were analyzed and presented.

3.3 RESULTS AND DISCUSSION

3.3.1 SUPPLEMENTARY IRRIGATION AND WATER USED

The data on effective rainfall, supplementary irrigation depth and water used is given in Table 3.1. The rainfall during the cropping season was 336 and 320 mm. The effective rainfall was 269 and 256 mm and the water harvested in the farm pond was 1350 m^3 and 920 m^3 during 2013–14 and 2014–15, respectively.

TABLE 3.1 Effect of Treatments on Supplementary Irrigation (SI) Depth and Water Used

Treatment	Effective rainfall (mm)		Supplementary irrigation depth (mm)		Water used (mm)	
	2013–14	2014–15	2013–14	2014–15	2013–14	2014–15
No SI	269	256	0	0	269	256
SI – once	269	256	30	20	299	276
SI – twice	269	256	50	40	319	296
SI – thrice	269	256	70	60	339	316

The irrigation depth was 30, 50 and 70 mm for one, two, and three supplementary irrigation during 2013–14; and 20, 40 and 60 mm for one, two, and three supplementary irrigation during 2014–15. The water used in control without irrigation, one, two, and three supplementary irrigation treatments was 269, 299, 319 and 339 mm during 2013–14 and 256, 276, 296 and 316 mm during 2014–15. The water used was higher with three supplementary irrigations during both the years (339 and 316 mm). The water used in three supplementary irrigated plots was 5.8, 11.7 and 20.6% during 2013–14 and 7.8, 15.6 and 23.4% during 2014–15; these values were higher compared to two, one, and no supplementary irrigation plots. Results agreed with those reported by Oweis et al. (1996) and Sandhu et al. (1995).

3.3.2　CROP YIELD

During both the years of experimentation, minimum tillage + crop residue and supplementary irrigation had a significant influence on grain yield of black gram and was presented in Table 3.2.

TABLE 3.2　Effect of Tillage + Crop Residue and Supplementary Irrigation on Yield of Black Gram (kg ha^{-1})

Treat-ments	Back gram yield, kg/ha									
	2013–14					2014–15				
	Tillage + Crop Residue					Tillage + Crop Residue				
	M1	M2	M3	M4	Mean	M1	M2	M3	M4	Mean
	Supplement Irrigation, SI									
S1	595	612	786	796	697	528	565	726	735	639
S2	631	737	838	967	793	583	681	774	893	733
S3	719	798	870	986	843	664	731	804	911	779
S4	292	398	526	604	455	256	367	492	539	414
Mean	559	636	755	838	—	508	588	699	770	—
	M	S	M at S	S at M	—	M	S	M at S	S at M	—
SEd	31	23	19	55	—	26	17	18	41	—
CD (5%)	76	47	43	94	—	64	34	27	56	—

Minimum tillage + crop residue @ 5 t ha^{-1} gave higher yield (838 kg ha^{-1} and 770 kg ha^{-1}) and followed by minimum tillage + crop residue @ 2.5 t ha^{-1}. Significantly lower yield was recorded by conventional tillage. The results agreed with Fisher et al. (2002). Three supplementary irrigations through rain-gun sprinklers (843 kg ha^{-1} and 779 kg ha^{-1}) enhanced the yield compared to two supplementary irrigations through rain-gun sprinklers (793 kg ha^{-1} and 733 kg ha^{-1}) during both years. Giving three supplementary irrigations through rain-gun sprinklers from rainwater in pond boosted the yield up to 46% in black gram, compared to black gram grown without supplementary irrigation as indicated by Wong et al. (2013).

The interaction effect was significant during both the years. Minimum tillage + crop residue @ 5 t ha^{-1} with three supplementary irrigations through rain-gun sprinklers registered higher black gram grain yield (986 kg ha^{-1} and 911 kg ha^{-1}) and was comparable with the combined effect of minimum tillage + crop residue @ 5 t ha^{-1} with two supplementary irrigations through rain-gun sprinklers (967 kg ha^{-1} and 893 kg ha^{-1}). The combined effect of conventional tillage and no supplementary irrigation drastically reduced the yield. Minimum tillage + crop residue with supplementary irrigation increased the soil moisture content and reduced the stress during the critical period and thereby resulted in higher yield as reported by Anwar et al. (2004) and Govaerts et al. (2005).

3.4 SUMMARY

Unbalanced rainfall distribution and water deficiency are major problems threatening dryland agriculture in Tamil Nadu. Field experiments were conducted to identify technologies, which can sustain productivity with limited water in drylands. Rainwater was harvested in the farm pond and same was used for supplementary irrigation. Minimum tillage + crop residue @ 5 t ha^{-1} with two or three supplementary irrigations of 40–60 mm using rain-gun sprinklers enhanced the productivity of black gram. Therefore, this recommendation is extended for sustaining productivity in drylands of Tamil Nadu.

KEYWORDS

- conservation agriculture
- dryland productivity
- farm ponds
- rain-gun sprinklers
- supplementary irrigation
- water use

REFERENCES

Anwar, S., Khaliq, A., Nabi, G., & Zafar, M., (2004). Use of rain-gun sprinkler system for enhancement of wheat production. *Pakistan Journal of Life and Social Sciences, 2*(2), 174–177.

Dastane, N. S., (1974). *Effective Rainfall*. Irrigation drainage paper 25, Food and Agriculture Organization (FAO), Rome, p. 69.

Delgado, J. A., (2010). Crop residue is a key for sustaining maximum food production and for conservation of our biosphere. *Journal of Soil and Water Conservation, 65*(5), 111–116.

Fischer, R. A., Santiveri, F., & Vidal, I. R., (2002). Crop rotation, tillage and crop residue management for wheat and maize system performance in the sub humid tropical highlands. *Field Crops Research, 79,* 123–137.

Govaerts, B., Sayre, K. D., & Deckers, J., (2005). Stable high yields with zero tillage and permanent bed planting. *Fields Crop Research, 94,* 33–42.

Krishna, J. H., Arkin, F. G., & Martin, J. R., (1987). Runoff impoundment for supplemental irrigation in Texas. *Journal of the American Water Resource Association, 23*(6), 1057–1061.

Mostafa, H., & Derbala, A., (2011). Performance of supplementary irrigation systems for corn silage in the sub-humid areas. *Agricultural Engineering International, 15*(4), 9–15.

Oweis, T., & Taimeh, A., (1996). Evaluation of a small water harvesting system in the arid region of Jordan. *Water Resources Management, 10,* 21–34.

Oweis, T., (1997). *Supplemental Irrigation: A Highly Efficient Water-Use Practice.* ICARDA, Aleppo, Syria, p. 16.

Sandhu, K. S., & Sandhu, A. S., (1995). Response of dryland wheat to supplemental irrigation and rate and method of N application. *Nutrient Cycling in Agroecosystems, 45*(2), 135–142.

Somasundaram, S., & Avudaithai, S., (2011). Effect of rain-gun sprinkler irrigation on different crops in alkaline environment. *Green Farming, 2*(2), 196–198.

Wang, D., (2017). Water use efficiency and optimal supplemental irrigation in a high yield wheat field. *Field Crops Research, 155,* 213–220.

Wang, D., Yu, Z. W., & White, P. J., (2013). The effect of supplemental irrigation after jointing on leaf senescence and grain filling in wheat. *Field Crops Research, 151,* 35–44.

CHAPTER 4

IMPACT OF IRRIGATION METHODS ON SOIL, WATER, AND NUTRIENT USE EFFICIENCY OF INTEGRATED CROPPING: LIVESTOCK PRODUCTION SYSTEMS

V. S. MYNAVATHI, R. MURUGESWARI,
V. RAMESH SARAVANA KUMAR, H. GOPI,
C. VALLI, and M. BABU

ABSTRACT

Climate change may adversely affect various aspects of livestock production systems including animal health and productivity, fodder production, water availability, pest, and diseases. Water applied using these systems supports the growth of annual food crop rice and perennial fodders, yielding a cost-effective production system. The research was conducted to study the impact of irrigation methods (Furrow vs. Drip in Fodder crops, Flood irrigation vs. SRI system in Rice) on the productivity of Rice, Napier hybrid grass and *Desmanthus* nutritious fodder species in Kancheepuram District of Tamil Nadu. Alternate wetting and drying irrigation reported to save water compared with continuous flooding in rice cultivation. Water use efficiency indicated that the utilization of water for every kg of crop production and green fodder was reduced by introducing the SRI cultivation of rice and drip irrigation in green fodder cultivation. The water saving was 37.5% in the paddy field. Irrigation method impacted green biomass yield (higher with furrow irrigation) but both methods yielded similar dry biomass. Results revealed that the controlled application of water through drip irrigation is able to produce more quantity of green fodder, leading to more effective utilization and resource conservation of available land, fertilizer, and water. Higher productivity of these nutritional fodders resulted in higher milk productivity of livestock. The ability to grow fodder crops year-round with limited water and water efficient drip irrigation may greatly increase livestock productivity and, hence, the economic security of livestock farmers.

4.1   INTRODUCTION

Improved irrigation use efficiency is an important tool for intensifying and diversifying agriculture, resulting in higher economic yield from irrigated farmlands with a minimum input of water. Research was conducted to study the impact of irrigation method (Furrow vs. Drip in Fodder crops; Flood irrigation vs. SRI system in Rice) on the productivity of Rice, Cumbu Napier hybrid grass and *Desmanthus* nutritious fodder species in Kancheepuram District of Tamil Nadu. The background of this research project is to preserve the soil nutrients like nitrogen, phosphorus, potassium, and organic carbon to the same soil through soil water efficient utilization in the crop (Paddy, fodder Cumbu Napier hybrid and *Desmanthus*), integrating with livestock (cows and goats). The scheme was implemented in the institute's land at Kancheepuram District to assess the effective utilization of soil nutrients.

4.2   MATERIALS AND METHODS

Nine dairy cows and nine goats were selected at institutional land. Based on the feeding requirement of livestock, fodders are produced. Soil profile was analyzed from the field of paddy for $Co(CN)_4$ and Desmanthus before start the study. Paddy crop and perennial fodder were grown in the required land area. Standard management practices were adopted to cultivate rice and fodder. A split-block factorial design was used. The factors considered were treatment location, fodder crop, and irrigation methods. Commonly used local agronomical practices were followed in all cases except irrigation method.

The crop residue of paddy straw was fed to cows. Green fodder was fed to cows and goats. The quantity of straw and green fodder fed to the animals was recorded. Animals were allowed for grazing in paddy-harvested land. The animal residue (dung and urine) were collected and measured. The collected dung was stored and recycled to the same land where crop and fodder were grown for feeding these animals. Crop yield, crop residue yield, fodder yield, production, and reproduction parameters of animals integrated with cropping system were recorded.

The soil nutrients including organic carbon in this project land are utilized through the production of crops, crop residue and fodder fed. These soil nutrients are conserved through recycling dung and urine by integrating animals in this research program. The production and reproduction performance of the animals are also enhanced by utilizing the organic farm producing crop residues and green fodder.

4.3 RESULTS AND DISCUSSION

4.3.1 WATER USE EFFICIENCY

The objectives of this research study to evaluate the impact of soil, water, and nutrients use efficiency on integrated crop and livestock production system in the Kancheepuram district of Tamil Nadu.

The quantity of water utilized for Paddy and green fodder cultivation for two years is presented in Table 4.1. Alternate wetting and drying method of irrigation saved water in rice cultivation. Optimum supply of irrigation water with mechanical weeding resulted in higher nutrient availability, subsequently resulting in the better source to sink conversion, which in turn enhanced production of more number of spikelets and filled grains panicle^{-1}. Similar findings were reported by Ancy (2007).

TABLE 4.1 Water Utilization for Paddy and Green Fodder Cultivation in Different Agro Zones

Parameters	End of 1st year	End of 2nd year
	Paddy	
Cultivation method	SRI	SRI
Water utilized (lit/kg)	2500	2500
Water savings (%)	37.5	37.5
	Co(CN)$_4$ and *Desmanthus*	
Drip irrigation	No	Yes
Water utilized (lit/kg)	50	38
Water savings (%)	-	24

4.3.2 IMPACT OF SOIL NUTRIENTS AND WATER UTILIZED

The data on the impact of soil nutrient utilized (%) and water utilized (%) are furnished in Table 4.2. Soil nutrient utilized (%) and water utilized (%) varied due to water, fertilizers, and manure applied to the crop in each location of the study area.

With regard to soil nutrient utilized of the study area, Northeastern zone recorded 100% of soil nutrient utilized during 2013–14 and 2014–15. During the year 2014–15, paddy cultivation recorded 62.5% water use. Nutrients may be very effective when applied continuously through the irrigation system at rates not exceeding the requirements of the plants, which is in agreement with the results by Bar-Yosef and Sagiv (1982).

In paddy cultivation, alternate wetting and drying method of irrigation resulted in higher nutrient use efficiency compared to conventional planting. These findings were in accordance with the findings by Debashis et al. (1999). Though the authors applied the entire quantity of nutrients by fertilizers, the crop uptake was low and therefore the nutrients were not utilized efficiently. However, through recycling of animal manures, the nutrients were supplied according to the need of the crop and dosage increased according to the stages and thus the applied nutrients were utilized efficiently.

The nutrient use efficiency was lower because the crop uptake could increase as the dosage increases and then slows down or declines after a critical limit. The excess dose of fertilizer might have resulted in high leaching loss and denitrification of applied fertilizers though the yield was high under excess fertilizers level. This reveals that applying excess fertilizer leads to soil pollution due to the leaching of nutrients. Thus the optimum dose of nutrients was sufficient to achieve higher yield by reducing the leaching of the nutrients and preventing the soil pollution to some extent.

Better management of water by frequent application of small, calculated amounts during the growth cycle helps to maintain N in the root zone and improves N efficiency.

Scheduling fertilizer application based on the need of the crop once in three days offered the possibility of reducing nutrient losses, thereby increased the nutrient use efficiency when compared to conventional application methods.

TABLE 4.2 Impact of Soil Nutrient Utilized (%) and Water Use (%)

Zones	Soil nutrient utilized (%)*						Water utilized (%)			
Year	2013–14			2014–15			2013–14		2014–15	
Crop/Fodder	Paddy	Co(CN)4	Des	Paddy	Co(CN)4	Des	Paddy	Co(CN)4 & Desmanthus	Paddy	Co(CN)4 & Desmanthus
Northeastern zone*	100	100	100	100	100	100	62.5	100	62.5	76

*No fertilizer applied

The increase in water utilized over the conventional system was mainly due to a considerable saving of irrigation water, a greater increase in crop yield and higher nutrient use efficiency. This was in concordance with findings by Suhas et al. (2002) and Ramah (2008). Increase in irrigation amount did not increase the marketable yield of crops but reduced the irrigation production efficiency significantly (Imtiyaz et al., 2000). Ardell (2006) reported that application of N and P fertilizer will frequently increase crop yields, thus increasing crop water use efficiency. Adequate levels of essential plant nutrients are needed to optimize crop yields and water utilized.

The lower water utilized under surface irrigation might be due to higher consumption of water and lower yield recorded. Similar results were obtained by Ahluwalia et al. (1993).

4.4   SUMMARY

Results revealed that location effect was significant ($p < 0.01$) with highest fodder productivity. Species effects were also significant, with *Cumbu Napier* hybrid grass having a higher yield than *Desmanthus*. Irrigation method impacted green biomass yield (higher with furrow irrigation) but both methods yielded similar dry biomass, while water use was 73% less under drip irrigation. The results revealed that the controlled application of water through drip irrigation is able to produce more quantity of green fodder, and the controlled water use leads to more effective utilization and resource conservation of available land, fertilizer, and water. Higher productivity of these nutritional fodders resulted in higher milk productivity of livestock.

KEYWORDS

- **drip irrigation**
- **forage biomass**
- **irrigation efficiency**
- **Napier grass**

REFERENCES

Ahluwalia, M. S., Bladev, S., & Gill, B. S., (1993). Drip irrigation system-its hydraulic performance and influence on tomato and cauliflower crops. *J. Water Manage*, *1*, 6–9.

Ancy, F., (2007). *Evaluation of Different Crop Establishment Methods for Increasing Yield in Transplanted Hybrid Rice*. M.Sc. (Ag.) Thesis for Tamil Nadu Agricultural University, Coimbatore, Tamil Nadu, India, p. 130.

Ardell, D. H., (2006). Water use efficiency under different cropping situation. *Ann. Agric. Res.*, *27*(5), 115–118.

Bar-Yosef, B., & Sagiv, B., (1982). Response of tomatoes to N and water applied via a trickle irrigation system. I. Nitrogen. *Agron. J.*, *74*, 633–639.

Debashis, C., Singh, A. K., Kumar, A., & Khanna, M., (1999). Movement and distribution of water and nitrogen in soil as influenced by fertigation in broccoli (*Brassica oleracea* var. *italica* L.). *J. Water Management*, *7*(1&2), 8–13.

Imtiyaz, M., Chepet, K., & Mothobi, E O., (2000a). Yield and economic return of vegetables crops under variables irrigation. *Irrig. Sci.*, 1987–1993.

Ramah, K., (2008). Study on Drip Irrigation in Maize (*Zea mays* L.) Based Cropping System. PhD Thesis for TNAU, Coimbatore, p. 124.

Suhas-Bobade, V., Asokaraja, N., & Arthanari, P. M., (2002). Effect of drip irrigation and nitrogen levels on yield, water use and water use efficiency of eggplant. *Crop Res.*, *24*(3), 481–486.

CHAPTER 5

ESTIMATION OF RESOURCE USE PATTERNS OF CROPS CULTIVATED ACROSS DIFFERENT FARM SIZES

A. REVATHY and D. DAVID RAJASEKAR

ABSTRACT

The present study analyzes the resource use pattern of crops cultivated across different size groups of farms in Tamirabarani irrigation system. Tamil Nadu has 17 major river basins, which are mostly water stressed. Any strategy towards the improvement in total food grain production cannot omit these potential river basins. With this in view, the present study has been undertaken in the Tamirabarani irrigation system in Tirunelveli district to explore the efficiency of crop production. There are 133 system tanks and seven channels in Tirunelveli district commanding a total area of 16161 ha of which 11031 ha comes under direct command area and the rest 5130 ha is under indirect irrigation, where the canal water is fed into system tanks and then used for irrigation indirectly. Coefficient Variation was worked out to study the variation in the farms across the scale of farming in different reaches. A three-stage random sampling procedure was used to select 180 sample respondents. The field enquiry was made during January to March of 2014. The resource use pattern analysis revealed that the productivity and the usage of different resources across small, medium, and large farms in the production of paddy and banana were higher in the head region than the mid region. Strengthening of extension efforts towards sensitizing all farmers about the technology packages for different crops by appropriate training and, demonstration, and in particular to the different size group of farmers in mid-region of the irrigation system.

5.1   INTRODUCTION

India is endowed with 196.9 million ha of arable land and it holds second largest agricultural land area in the world. Agriculture accounts for 12% in national GDP and 11% in the total export earnings and it also provides raw material to a large number of industries such as textiles, silk, sugar, rice, flourmills, and milk products. According to 2012 statistics, the total food grain production of India was 259.32 million tons. The growth in the production of agricultural crops depends on many factors such as area cropped, input management and yield. The cropped area and productivity are determined by the fertility of soil, monsoon pattern, rainfall, irrigation, availability of agricultural laborers, climatic changes and agricultural prices. Groundwater, one of the major sources of irrigation in India, is being rapidly depleted. Out of total 5700 blocks existing in India, the number of dark blocks, where groundwater extraction is more than 85% of the availability, has increased from 253 to 428 between 1984–85 and 1998–99 (GOI, 2002).

It is estimated that Tamil Nadu may require total food grain production of 12.32 million tons in the year 2020 to meet the consumption demand alone without considering the industrial demand; while the production during 2010–11 was only about 7.59 million tons, leaving a large demand and supply gap of 4.76 million tons. Bridging this wide gap calls for expanding the cropping area, improving cropping intensity and bringing about an appropriate shift in cropping pattern in different districts of Tamil Nadu (GOI, 2002, 2012). Therefore, development and promotion of sustainable farm planning models and their management based on scientific principles seems to be the probable approach to address the problem of sustainable production and to meet the economic and social objectives of its ever-growing population and to satisfy the ever-expanding needs

Tamil Nadu has an area of 13 million ha of which net area sown area constitutes 37.5%. Irrigated crops account for 56.80% of Gross Cropped Area and the rest 43.2% is under rain-fed crops. Irrigation is an important input needed for agricultural development and it is essential for the adoption of green revolution technologies such as high yielding varieties, chemical fertilizers, and pesticides. Source wise irrigation indicated that open wells are the main source of irrigation, which accounted for 1.570 million ha, followed by canals with 0.747 million ha, tanks with 0.533 million ha, tube wells with 0.403 million ha and bore-wells and other sources with 9,068 ha.

The Tamirabarani irrigation system is one of the oldest irrigation systems in Tamil Nadu. The river with its tributaries having their origin in Western Ghats from its source in the Pothigai hills drains into the Gulf of Mannar at Punnakkayal village near Tiruchendur. The river originates in Tirunelveli district and confluences with the Bay of Bengal in Thoothukudi district, traversing a length of about 120 km. Some of the tributaries of this river are Servalar, Manimuthar, Varahanadhi, Pachayar, and Chittar. Run-off occurs both during the South West and North East monsoon periods, thus making it a perennial river. In Kautiliya'sArthasastra, it is referred as to as 'Parasamudra'. Eight dams have been constructed across the main river till date (six dams in Tirunelveli and two in Thoothukudi districts). There are in total eleven channels served by these eight dams (seven channels comes under Tirunelveli district and four under Thoothukudi district) (CARDS, 2008).

5.2   MATERIALS AND METHODS

5.2.1   STUDY AREA AND DATA

The head and middle region of Tamirabarani river basin covered under Tirunelveli district formed the basis of this study. A three-stage random sampling method was adopted to select the sample farmers:

- At the first stage, the blocks of Tirunelveli district is arranged in ascending order based on the gross cropped area and three blocks were selected.
- At second stage, all the revenue villages from each of the selected blocks were arranged in ascending order based on the gross cropped area in the year 2011–12 and three revenue villages per block were selected, thus constituting 12 selected revenue villages with the spread of six revenue villages each in head and middle reaches, respectively.
- Using the same criterion, all the farmers in each of the selected 12 revenue villages were arranged in the ascending order separately, and 15 farmers were selected from each of the 12 selected revenue villages, thus constituting a total sample size of 180 farmers with a spread of 90 farmers each in head and middle reaches of the river basin, respectively.

The survey was taken in the agricultural season during January to March of 2014. The selected sample respondents were personally contacted and the required primary data were collected through personal interview method by using the pre-tested interview schedule. Cropping pattern analysis in head and mid-region of Tamirabarani river basin revealed that paddy was the predominant crop in all farms, followed by banana, bhendi, brinjal, black gram, chilly, tomato, and red gram (TNAU).

5.2.2   COEFFICIENT OF VARIATION

Analysis of variance was conducted to test the homogeneity of head and middle region of Tamirabarani irrigation system with respect to the gross cropped area of blocks and villages and found that significant difference exists between head and mid-region while there was no significant difference existing between the villages in the head region as well as in the mid-region of the Tamirabarani irrigation system, respectively.

Coefficient Variation of the selected variables such as crop expenses, crop income, consumption expenditure, the yield of crops, the value of assets and resource endowments were worked out to study the variation in the above-said variables in the farms across the scale of farming in different reaches of Tamirabarani river basin.

5.3   RESULTS AND DISCUSSION

5.3.1   RESOURCE USE PATTERN OF PADDY IN SMALL FARMS OF HEAD AND MID REGION

The average production of paddy per ha in small farms of the head region was 2860.54 kg (TNAU) and it was more than the average productivity of mid-region with 2817 kg and the average productivity of both regions was lesser than the standard paddy yield of 4750 kg/ha (Table 5.1).

The comparative analysis of average input usage for paddy with the standard recommended dosage in small farms across regions indicated that the average input usage in both regions was found to be lesser than standard usage in seed, farmyard manure, machine use, phosphorus, potash, which indicated the scarcity of these inputs. Among the scarce inputs, farmyard manure and machine use exhibited very high scarcity. The comparative analysis of input usage in small farms across regions indicated that usage of seeds, nitrogen, phosphorus, potash, and water was found higher in the head region than in mid region, which might be the reason for higher yield of paddy in the head region than in mid region. The consistency in usage of inputs such as seed, farmyard manure, nitrogen, and water was better in the small farms of the head region than mid region.

TABLE 5.1   Resource Use Pattern of Paddy in Small Farms of Head and Mid-Region

Variables	Standard Requirement [5]	Head Region		Mid Region	
		Mean	CV	Mean	CV
Seed in kg/ha	70	40.05	38.42	39.93	40.33
Farmyard manure in tons/ha	12.50	5.33	49.76	5.36	52.42
Labor in man-days/ha	160	141.55	6.00	142.01	7.35
Machine hours in hrs/ha	12.50	5.97	29.65	6.05	29.55
Nitrogen in kg/ha	150	145.05	7.15	140.80	8.81
Phosphorus in kg/ha	60	48.50	9.87	31.32	8.27
Potash in kg/ha	60	47.75	10.91	32.55	7.58
Irrigation in Ha mm	1200	1075.00	3.20	1057.70	4.36
Plant Protection in	950	743.75	36.80	757.91	35.09
Rs/ha (US$/ha)	(15.83)	(12.40)	(0.63)	(12.60)	(0.58)
Yield in kg/ha	4750	2860.54	32.48	2817.00	32.98

Note: In this chapter: Rs. 60.00 = 1.00 US$.

5.3.2   RESOURCE USE PATTERN OF PADDY IN MEDIUM FARMS OF HEAD AND MID REGION

The average production of paddy per ha in medium farms of the head region was 4633.89 kg and it was more than the average productivity of mid-region with 4190 kg and average productivity of both regions was lesser than the standard paddy yield of 4750 kg/ha (Table 5.2).

The comparative analysis of average input usage for paddy with the standard recommended dosage in medium farms across regions indicated that the average input usage in both regions was found to be lesser than standard dosage in farmyard manure, machine use, phosphorus, potash, and nitrogen, which indicated the scarcity of these inputs. Among scarce inputs, farmyard manure and machine use exhibited very high scarcity.

The comparative analysis of input usage in medium farms across regions indicated that usage of seeds, nitrogen, labor, potash, and water was found higher in the head region than in mid region, which might be the reason for higher yield of paddy in the head region than in mid region. The consistency in usage of inputs such as seed, labor, machine hours, phosphorus, and water was better in the head region than in mid region.

TABLE 5.2   Resource Use Pattern of Paddy in Medium Farms of Head and Mid Region

Variables	Standard Requirement	Head Region		Mid Region	
		Mean	CV	Mean	CV
Seed in kg/ha	70	69.08	10.35	65.96	16.78
Farmyard manure in tons/ha	12.50	8.50	38.07	8.70	23.19
Labor in man days/ha	160	148.53	2.89	117.39	8.32
Machine hours in hrs/ha	12.50	7.65	14.57	7.91	18.22
Nitrogen in kg/ha	150	128.15	5.28	106.23	4.78
Phosphorus in kg/ha	60	37.45	8.92	38.38	10.05
Potash in kg/ha	60	34.91	26.01	33.57	27.57
Irrigation in Ha mm	1200	1121.16	9.37	1020.89	9.62
Plant protection in	950	487.50	18.26	600.50	20.50
Rs/ha (US$/ha)	(15.83)	(8.10)	(0.30)	(10.00)	(0.34)
Yield in kg/ha	4750	4633.89	3.73	4190.00	23.35

5.3.3   RESOURCE USE PATTERN OF PADDY IN LARGE FARMS OF HEAD AND MID-REGION

The average per ha production of paddy in large farms of the head region was 2870 kg and it was less than the average productivity in mid region with 3778.00 kg and the average productivity in both regions was lesser than standard paddy yield of 4750 kg/ha (Table 5.3).

The comparative analysis of average input usage for paddy in large farms across regions with standard recommended dosage indicated that the average usage of inputs in both regions was found to be lesser than standard dosage in farmyard manure, machine use, nitrogen, phosphorus, potash, which indicated the relative scarcity of these inputs. Among the scarce inputs, farmyard manure and machine use exhibited very high scarcity. The comparative analysis of input usage between head and mid-region indicated that usage seeds, farmyard manure, machine usage, nitrogen, phosphorus, potash, and water was found more in the head region than in mid region, which might be the reason for higher yield of paddy in the head region than in mid region. The consistency in usage of seeds, farmyard manure and water was better in the large farms of the head region than in mid region.

TABLE 5.3   Resource Use Pattern of Paddy in Large Farms of Head and Mid Region

Variables	Standard Requirement	Head Region		Mid Region	
		Mean	CV	Mean	CV
Seed in kg/ha	70	67.77	2.63	62. 72	2.98
Farmyard manure in tons/ha	12.50	6.50	35.40	6.32	38.05
Labor in man days/ha	160	106.77	19.04	102.92	19.51
Machine hours in hrs/ha	12.50	9.37	18.37	9.00	8.95
Nitrogen in kg/ha	150	116.84	16.59	115.11	16.79
Phosphorus in kg/ha	60	43.25	10.29	42.56	11.13
Potash in kg/ha	60	45.90	16.43	45.31	16.60
Irrigation in Ha mm	1200	1127.56	12.00	1026.75	29.32
Plant protection in	950	535.41	15.23	723.75	13.57
Rs/ha (US$/ha)	(15.83)	(8.92)	(0.25)	(12.06)	(0.23)
Yield in kg/ha	4750	2870.00	31.98	3778.00	16.27

5.3.4   RESOURCE USE PATTERN OF BANANA IN SMALL FARMS OF HEAD AND MID REGION

The average productivity of banana in small farms of the head region was 21614.20 kg/ha and it was more than the average productivity of mid-region with 20477.50 kg and the average productivity in both regions was lesser than the standard banana yield of 25000 kg/ha (Table 5.4). The comparative analysis of average input usage for the banana in small farms across regions with standard recommended dosage indicated that the average usage of inputs for the banana in both regions was found to be lesser than standard requirement in inputs such as seed, farmyard manure, machine use, phosphorus, which indicated the scarcity of these inputs. Among these scarce inputs, farmyard manure and plant protection chemicals exhibited very high scarcity. The comparative analysis of input usage in small farms between in head and mid region with respect to banana indicated that usage of labor, nitrogen, and plant protection chemicals was found higher in the head region than in mid region, which might be the reason for higher yield of banana in the head region than in mid region. The consistency in the usage of labor and machine power was better in the small farms of the head region than in mid region.

TABLE 5.4   Resource Use Pattern of Banana in Small Farms of Head and Mid Region

Variables	Standard Requirement [4]	Head Region		Middle Region	
		Mean	CV	Mean	CV
Seed suckers in No/ha	2500	2047.18	4.97	2141.50	3.99
Farmyard manure in tons/ha	25	7.29	58.66	8.19	45.84
Labor in man days/ha	512	404.08	2.14	333.05	8.17
Machine hours in hrs/ha	15	11.55	12.29	13.24	16.52
Nitrogen in kg/ha	150	125.48	33.04	121.80	26.09
Phosphorus in kg/ha	100	59.78	40.83	58.91	31.50
Potash in kg/ha	200	198.56	16.65	201.38	15.43
Irrigation Ha mm	2200	1121.31	24.20	1022.00	23.18
Plant Protection in Rs/ha (US$/ha)	1500	1106.98	52.78	833.06	39.37
Yield in kg/ha	25000	21614. 20	8.77	20477.50	10.63

5.3.5   RESOURCE USE PATTERN OF BANANA IN MEDIUM FARMS OF HEAD AND MID REGION

The average productivity of banana in medium farms of the head region was 17882.30 kg/ha and it was more than the average productivity in mid region with 16485.20 kg and the average productivity in both regions was lesser than the standard banana yield of 25000 kg/ha (Table 5.5). The comparative analysis of average input usage in the head and mid region with the standard recommended dosage indicated that the average input usage for the banana in both regions was found lesser than the standard recommended dosage in inputs such as farmyard manure, machine use, labor, which indicated the scarcity of these inputs. Among these scarce inputs, farmyard manure and machine use exhibited very high scarcity. The comparative analysis of input usage across regions indicated that usage of nitrogen, phosphorus, potash, farmyard manure and plant protection chemicals was found higher in the head region than in mid region, which might be the reason for higher yield of banana in the head region than in mid region. The consistency in the usage of inputs such as potash, irrigation, farmyard manure and machine hours was better in the medium farms of the head region than in mid region.

TABLE 5.5 Resource Use Pattern of Banana in Medium Farms of Head and Mid Region

Variables	Standard Requirement	Head Region		Middle Region	
		Mean	CV	Mean	CV
Seed suckers in No/ha	2500	2059.17	8.47	2424.18	3.40
Farmyard manure in tons/ha	25	4.87	39.33	3.90	45.39
Labor in Man days/ha	512	342.23	9.55	432.12	6.59
Machine hours in hrs/ha	15	7.14	37.11	6.64	49.60
Nitrogen in kg/ha	150	209.19	32.43	182.43	6.20
Phosphorus in kg/ha	100	109.69	25.24	91.27	3.30
Potash in kg/ha	200	231	14.65	210	15.56
Irrigation Ha mm	2200	1126.82	6.07	1025.37	15.70
Plant Protection in	1500	1545.56	38.74	1318.12	15.79
Rs/ha (US$/ha)	(25.00)	(25.75)	(0.65)	(21.97)	(0.26)
Yield in kg/ha	25000	17882.30	23.76	16485.20	35.00

5.3.6 RESOURCE USE PATTERN OF BANANA IN LARGE FARMS OF HEAD AND MID REGION

The average productivity of banana in large farms of the head region was 19,943.30 kg /ha and it higher than the average productivity in mid region with 194,200 kg and the average productivity in both regions was lesser than the standard banana yield of 25,000 kg/ha (Table 5.6). The comparative analysis of average input usage in the head and mid region with the standard recommended dosage indicated that the average input usage for the banana in both regions was found lesser than the standard recommended dosage in inputs such as seed, farmyard manure, machine use, which indicated the scarcity of these inputs. Among these scarce inputs, farmyard manure and machine use exhibited very high scarcity. The comparative analysis of input usage across regions indicated that usage of suckers, potash, and plant protection chemicals was found higher in the head region than in mid region and the higher dose potash in the head region which might be the reason for higher yield of banana in the head region than in mid region. The consistency in the usage of inputs such as potash, irrigation, farmyard manure and machine hours was better in the medium farms of the head region than in mid region.

TABLE 5.6   Resource Use Pattern of Banana in Large Farms of Head and Mid Region

Variables	Standard Requirement	Head Region		Middle Region	
		Mean	CV	Mean	CV
Seed Suckers in No/ha	2500	1646.33	25.17	1608.66	29.32
Farmyard manure in tons/ha	25	6.25	25.25	6.47	26.00
Labor in man days/ha	512	353.80	6.11	358.60	3.60
Machine hours in hrs/ha	15	9.56	31.07	9.91	25.00
Nitrogen in kg/ha	150	242.18	19.97	250.42	26.00
Phosphorus in kg/ha	100	122.05	16.90	127.42	18.60
Potash in kg/ha	200	321.23	18.45	306.88	13.88
Irrigation Ha mm	2200	1228.80	25.82	1629.07	32.00
Plant Protection in	1500	1250.00	53.62	1223.00	34.80
Rs/ha (US$/ha)	(25.00)	(20.83)	(00.89)	(20.38)	(0.58)
Yield in kg/ha	25000	19943.30	34.77	19420.00	25.60

5.4   SUMMARY AND POLICY

The resource use pattern analysis revealed that the productivity and the usage of different resources across small, medium, and large farms in the production of paddy and banana were higher in the head region than the mid region. However, all the resource usage and the productivity of different crops in general was lesser than the standard recommendation and standard yield of the major crops in both regions of Tamirabarani river basin, which calls for strengthening extension efforts towards sensitizing the all farmers about the technology packages for different crops by appropriate training and, demonstration, and in particular to the different size group of farmers in mid-region of the irrigation system. Ensuring the availability of right machineries at block and village level and providing them on hire basis to farmers during the crop production season through cooperatives on the part of Government assumes importance to improve the efficiency of farming in the river basin. Apart, appropriate strategies and plans for the conversion of available farm wastes into vermicompost and farmyard manure at farmsteads to ensure an adequate supply of farmyard manure in the farms to conserve ecology on the one hand and to improve the efficiency of crop production by avoiding inefficient input mixes, on the other hand, are urgently needed.

KEYWORDS

- **coefficient of variation**
- **farm size**
- **resource use pattern**

REFERENCES

District Agriculture Plan Tirunelveli District, (2008). Centre for Agricultural and Rural Development Studies (CARDS), Tamil Nadu Agricultural University, Coimbatore, p. 54.

Government of India (GOI), (2002). Tenth Five Year Plan 2002–2007, Volume – II. Planning Commission Government of India, New Delhi, p. 201.

Government of India (GOI), (2012). Twelfth Five Year Plan. Planning Commission Government of India, New Delhi, p. 251.

Standard requirements for banana crops, (2004). In: *Crop Production Techniques of Horticultural Crops*. Directorate of horticulture and plantation crops, Tamil Nadu Agricultural University (TNAU), Chepauk, Chennai, p. 113.

Standard requirements for paddy crops, (2004). http://agritech.tnau.ac.in/agriculture/agri_nutrientmgt_rice.html. Accessed on July 31, 2017.

PART II
Water Use Efficiency for Drip Irrigated Fodder, Cassava, and Pearl Millet

WATER USE EFFICIENCY OF FODDER (*DESMANTHUS VIRGATUS*) UNDER CHECK BASIN, FURROW, AND RAIN-GUN IRRIGATION METHODS

V. S. MYNAVATHI, M. MURUGAN, and H. GOPI

ABSTRACT

The field experiment was conducted on bundle flower (*Desmanthus virgatus*) under check basin, furrow, and rain-gun systems at Post Graduate Research Institute in Animal Sciences, Kattupakkam – Tamil Nadu during 2013–2014. Irrigation efficiencies of different irrigation methods and yield of *Desmanthus virgatus* was compared. Water application and water use efficiency were maximum in case of rain-gun irrigation method compared to check basin and furrow irrigation. Use of rain-gun method of irrigation during the cropping period helped in water saving water when the soil infiltration rate was very high and water need in the root zone was lower. Using rain-gun irrigation system, 35% and 25% higher water use efficiency and 17.1% and 9.0% more water application efficiency were achieved compared to check basin and furrow irrigation method, respectively. About 3.2% and 1.52% higher yield was obtained in rain-gun irrigation system compared to check basin and furrow irrigation methods, respectively. Therefore, it was concluded that rain-gun irrigation method is a most feasible system for the production of fodder during water scarcity period. It reduces manpower requirement and prevents the growth of weeds, which flourish in check basin and furrow irrigation, thereby nutrient loss due to the utilization of nutrients by the weeds arising in flood irrigation will be minimized. Water is economically and efficiently used to produce maximum biomass with a minimum amount of water.

6.1   INTRODUCTION

Water application efficiencies at field level must be improved to overcome the shortage of water, to reduce the problem of waterlogging. It is, therefore, important to develop techniques to use available resources of irrigation water more efficiently during field application. Application efficiency can be increased by adopting pressurized irrigation systems like rain-gun irrigation, which is expensive to operate by small and marginal farmers. Studies have shown that well designed and well-managed surface irrigation systems have comparable application efficiencies to those of pressurized system. Therefore, it is important to improve surface irrigation systems and their management to increase application efficiency without lowering the biomass yield of fodder. Innovative technologies are needed to increase water use efficiency through the introduction of efficient irrigation systems. Irrigation technologies and irrigation scheduling should be adopted for more effective use.

This study was planned to determine the fodder yield of *Desmanthus virgatus* by using different irrigation methods and suitability of the most efficient system for green fodder production.

6.2   MATERIALS AND METHODS

The experiment was conducted at the farm of Post Graduate Research Institute in Animal Sciences (PGRIAS)located at Kattupakkam, Kancheepuram District of Tamil Nadu under Tamil Nadu Veterinary and Animal Sciences University, Chennai during 2013–2014. The soil was sandy loam. The 0.65 ha of the plot was divided into three portions for check basin, furrow, and rain-gun irrigation systems. The area for check basin and furrow was 0.20 ha each while area for rain-gun irrigation system was 0.25 ha. Three replications were used for each treatment. The size of the basin was 25 m x 25 m while in furrow irrigation system each replication consisted of 10 furrows with a length and width of 75 m and 0.75 m, respectively for each furrow. For rain-gun irrigation, 50 m x 50 m plot was selected and water was applied through rain-gun system.

Seeds of *Desmanthus virgatus* were sown in all irrigation plots using fertilizer rate of 10,60 and 30 kg/ha of N, P, and K, respectively. The seed was sown manually at a row-to-row distance of 0.75 m and continuous sowing within the row. Each irrigation was applied at 50% soil moisture deficit

(Michael, 1978) using measured quantity for basin and furrow irrigations, while for rain-gun irrigation the flow was measured from the storage tank for the specified time interval. The gravimetric method was used for soil moisture determination. The randomized complete block design was used.

6.3 RESULTS AND DISCUSSION

6.3.1 WATER USE EFFICIENCY

Higher water use efficiency of 27.5% was obtained with rain-gun irrigation system compared to check basin and furrow irrigation methods (Figure 6.1). The results indicated significant differences among the three irrigation systems. Higher water use efficiency of 0.85 kg/m^3 was obtained in rain-gun irrigation system compared to 0.6 kg/m^3 and 0.5 kg/m^3 in check basin and furrow irrigation systems, respectively. Similar findings were reported by Cetin and Bilgel (2002).

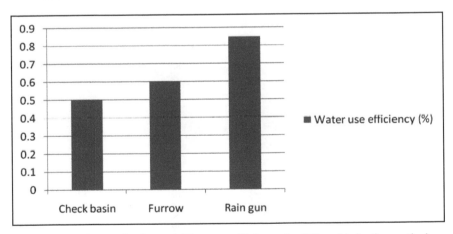

FIGURE 6.1   **(See color insert.)** Water use efficiency for different irrigation methods.

It was observed that the rain-gun irrigation system used the water more efficiently compared to other two irrigation systems. Water use efficiency in case of check basin and furrow irrigation system was nearly equal with only 0.1% difference, whereas this difference of efficiency was greater (30%) in case of the rain-gun irrigation system. Furthermore, the selection of rain-gun irrigation system depends upon the suitability of the system to socioeconomic conditions of the farmer, his technical skills and availability of servicing facilities and spare parts. The benefit-cost ratio of the rain-gun irrigation was found as 1.81, which indicated that the rain-gun sprinkler irrigation system was economically feasible.

6.3.2   WATER APPLICATION EFFICIENCY

The results of the water application efficiency for the check basin, furrow irrigation, and rain-gun irrigation are given in Figure 6.2, which showed highest application efficiency of 88% in case of the rain-gun irrigation system. The application efficiency of furrow irrigation system was 79%. Thus, by saving 25 mm depth of irrigation and using highest application efficiency of 88%, a reasonable increase in yield was achieved by rain-gun irrigation system. In addition to this, by achieving the highest application efficiency of 88% under Rain-Gun irrigation System.

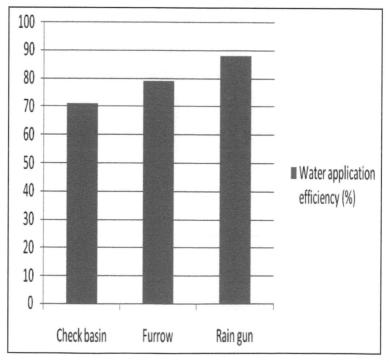

FIGURE 6.2   **(See color insert.)** Water application efficiency for three irrigation methods.

6.3.3   GREEN FODDER YIELD

It was observed that the green fodder yield in rain-gun irrigation system was 3.2% and 1.52% higher compared to check basin and furrow irrigation systems, respectively. It was concluded that by saving irrigation water, a reasonable increase in green fodder yield was obtained in the rain-gun irrigation system, for which green fodder yield of 112 tons/ha/year was obtained under rain-gun irrigation system.

6.4   CONCLUSIONS

- Water application and water use efficiencies were maximum in rain-gun irrigation system compared to basin and furrow irrigation system. Use of rain-gun irrigation during early crop season helped in water saving when the soil infiltration rate was high and the need of water for roots was less.
- About 35% and 25% higher water use efficiency was achieved by using rain-gun irrigation system compared to basin and furrow irrigation system, respectively.
- About 17.1% and 9.0% more water application efficiency were observed in using rain-gun irrigation system compared to basin and furrow irrigation systems, respectively.
- In the case of the rain-gun irrigation system, about 3.2% and 1.52% more grain yield was obtained compared to basin and furrow irrigation systems, respectively.

6.5   SUMMARY

Results revealed that well-designed and well-managed surface irrigation systems have comparable application efficiencies to those of pressurized system. Therefore, it is important to improve surface irrigation systems and their management to increase application efficiency without lowering the biomass yield of fodder. The study was planned to determine the fodder yield of *Desmanthus virgatus* by using different irrigation methods and suitability of the most efficient system for green fodder production. Use of rain-gun method of irrigation during the cropping period helped in water saving water when the soil infiltration rate was very high and the need of water in the root zone was less.

KEYWORDS

- efficiency
- fodder
- irrigation methods

REFERENCES

Cetin, O., & Bilgel, L., (2002). Effects of different irrigation methods on shedding and yield of cotton. *Agric. Water Management, 54,* 1–15.

Michael, A. M., (1978). *Irrigation Theory and Practice.* Vikas Publishing House Pvt. Ltd., 576 Masjid Road, New Delhi, p. 318.

CHAPTER 7

WATER PRODUCTIVITY OF MICRO-IRRIGATED CASSAVA (*MANIHOT ESCULENTA* CRANTZ)

SUNITHA SAROJINI AMMA and JAMES GEORGE

ABSTRACT

Cassava is a highly prospective crop among tropical tuber crops having food, feed, fuel, and industrial demand. Traditionally, cassava has been cultivated depending on the rainfall availability. Being a photo-insensitive crop, cassava can be grown throughout the year irrespective of the season, provided sufficient moisture is assured through irrigation. In the present-day context of climate change when precipitation has become irregular and scanty, it is essential to economize the use of irrigation water without compromising tuber yield. Field experiments were carried out in Kerala, India for three summer seasons, 2009–2010, 2010–2011 and 2011–2012 to investigate the response of cassava to micro irrigation. The treatments comprised of three levels of drip irrigation (I_1-Irrigation at 100% pan evaporation (PE), I_2–80% PE and I_3–60% PE) along with surface irrigation and a rainfed crop for comparison. Two node cuttings or Minisetts of variety Sree Vijaya (6 months) were planted during December every year, irrigation treatments were imposed, and data were collected on growth and yield parameters. Pooled data analysis indicated that irrigation at 100% PE resulted in maximum tuber yield (44 t ha^{-1}) and the benefit-cost ratio (4.27). Water productivity of cassava was 8.2 kg m^{-3} for I_1 level of irrigation compared to 4.2 kg m^{-3} under rainfed conditions and 2.6 kg m^{-3} under surface irrigation. On an average, water requirement of cassava was 3.0 mm per day during summer months.

7.1   INTRODUCTION

Cassava is the highly prospective crop among all the tropical tuber crops possessing food, feed, and industrial demand. Cassava has a major role as an industrial raw material for starch and sago production, which is taking place on a high-volume level in Tamil Nadu and has spread to adjacent states of Andhra Pradesh and Maharashtra. It is extensively grown in zones from latitude 30°N to 30S. Conventionally all tuber crops are cultivated depending on the rainfall availability. Even with improved varieties, advanced agricultural technologies and tools, the potential productivity of these crops are not attained, under rainfed conditions.

Cassava needs adequate moisture for sprouting and subsequent establishment of setts. It has a built-in mechanism to resist drought by shedding leaves and remaining dormant. When the rainfall starts, it draws on its root reserves to form new leaf to fill its roots (Ramanujam, 1994). However, earlier studies conducted in India revealed that when cassava is grown under rainfed conditions, supplementary irrigations during the drought period could give higher tuber yield than the rainfed crop (Nayar, et al., 1993). Though quite a drought tolerant, cassava produces better yield when regularly watered and the soil is not allowed to dry out completely (Mogaji-Kehinde, et al., 2011). Because of the increasing demand for the crop, presently, the cultivation has been extended to non-traditional areas with less rainfall, where the crop is mostly grown under surface method of irrigation in which major portion of irrigation water is lost by evaporation and deep percolation resulting in lower efficiencies.

Drip irrigation has proved to be a success in terms of water use efficiency in a wide range of crops. With drip irrigation, the soil is maintained continuously in a condition, which is highly favorable to crop growth (Edoga and Edoga, 2006). In the industrial belts of Tamil Nadu, farmers adopt drip irrigation, but without any rationale.

This chapter focuses on the response of cassava to micro irrigation; assessment of water requirement and water productivity of cassava under tropical conditions.

7.2   MATERIALS AND METHODS

Field experiments were conducted consecutively for three years 2009–10, 2010–11 and 2011–2012 at ICAR-Central Tuber Crops Research Institute,

Thiruvananthapuram – Kerala, India. The location lies between 8.54°N latitude and 76.91°E longitude and region comes under the humid tropical climatic zones of India at an elevation of 50 m above mean sea level. The region receives an average rainfall of 2100 mm, confined to mostly SW and NE monsoon seasons and the temperatures range between 24–32°C. Soil at the experimental site is kandiustult (Order: Ultisol) having pH in the acidic range, low in available nitrogen and medium in available phosphorus and potassium.

The study was conducted during December in all three years to make use of dry spell from December to May. Short duration variety (6–7months) of cassava developed by ICAR-CTCRI, Sree Vijaya was used for the study. Irrigation treatments included:

I_1 Irrigation at 100% of pan evaporation (PE);
I_2 Irrigation at 80 % PE; and
I_3 Irrigation at 60% PE.

The standard fertilizer dose (Kg per ha) of 100 N, 50 P_2O_5 and 100 K_2O was applied uniformly in all three treatments. Two control treatments, a rainfed crop as per recommended practices and surface flood irrigation at 5mm depth were also included for comparison.

Mini-setts of cassava (two node cuttings) were initially raised in a nursery during November. Seedlings with two to three fully opened leaves were uprooted and transplanted after four weeks. Transplanting was done at a spacing of 45 cm on ridges with a ridge to ridge spacing of60 cm. After the ridge formation, the drip system was laid out and drippers were placed to coincide with the spacing of the mini-setts. At the time of land preparation, Farm Yard Manure @ 12.5 t ha⁻¹ and a full dose of phosphorus fertilizer were applied as basal. The quantity of irrigation water in mm was calculated based on daily pan evaporation rate and pan factor. Crop factor was taken into account at different stages of growth as suggested by Allen and Pruitt (1991). Irrigation was given as per schedule from December to May. Nitrogen and potassium fertilizers were fertigated at weekly intervals up to 90 DA (days after planting). Biometric parameters (biomass production, yield attributes, and yield) were recorded. The crop was harvested by June. The data over the years were pooled and analyzed statistically following SAS procedure (**2010**). Water requirement for optimum production, water productivity, and economics were also determined.

7.3   RESULTS AND DISCUSSION

7.3.1   GROWTH AND GROWTH INDICES

Different irrigation treatments resulted in uniform plant height at different stages of growth except at 4 months after planting (MAP), where I_1 recorded the maximum height (40.4 cm). The rate of leaf production was similar at 2 MAP, but, varied significantly at later stages resulting in higher values at irrigation levels of 100% PE and 80% PE. Leaf area index (LAI) showed a marked gradation with decreasing levels of irrigation. The values were at par at 2 MAP, but I_1 resulted in maximum values both at 4 and 6 MAP (Figure 7.1), which was significantly higher than the other two irrigation levels. Crop growth rate (CGR) was maximum for irrigation at 100% PE during different intervals. The rate increased at a faster pace from 2 to 4 MAP under all levels of irrigation. All growth attributes were higher in the I_1 level of irrigation. Earlier findings (Nayar, et al., 1993) reported significantly greater dry matter production and CGR in cassava, which received supplementary irrigation.

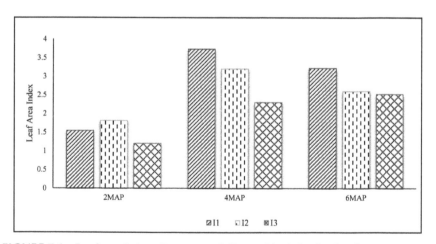

FIGURE 7.1   Leaf area index of cassava as influenced by irrigation levels.

7.3.2   YIELD ATTRIBUTES AND YIELD

All yield attributes and tuber yield were significantly affected by irrigation levels. Tuber bulking rate was more under I_2 and I_3 levels of irrigation during an initial period up to 4 months. However, during later stages of bulking, the rate of increase was higher under the I_1 level of irrigation. There was a sharp increase in tuber bulking rate after 4 months in I_1 compared to I_2 and I_3. There was a positive correlation between CGR and tuber bulking at three irrigation levels at different intervals. The average bulking rate was the highest with irrigation at 100% PE. In earlier studies also, tuber length and tuber girth of cassava showed a positive response to drip irrigation treatments indicating highest values with full irrigation (Khatib, et al., 2007; Mogaji-Kehinde, et al., 2011). A number of tubers produced per plant and harvest index did not vary significantly among all treatments. However, a higher tuber number was produced with 100% irrigation. Average tuber weight per plant was highest under I_1 (984 g), which was significantly superior to lower levels of irrigation.

Pooled data analysis revealed a declining trend in tuber yield with decreasing levels of irrigation (Table 7.1). Irrigation at 100% PE resulted in maximum tuber yield (43.94 t ha^{-1}), which was significantly superior over all other treatments. Irrigation at 60% PE was at par with surface irrigation. However, surface irrigation used almost 5 times more irrigation water compared to 60% in drip irrigation. Rainfed crop recorded the lowest tuber yield of 11 t ha^{-1}. During the summer season, in which the crop was raised for all three years, received an average of 230 mm effective rainfall only, and hence resulted in minimum yield. In cassava, maximum tuber yield of 36 t ha^{-1} was recorded with drip irrigation at 100% of surface irrigation followed by 75% of the surface irrigation and the values were at par (Amanullah, et al., 2006). In another irrigation experiment conducted in Nigeria, maximum dry matter and tuber yield were recorded when cassava was irrigated at a water regime of 100% available water (Odubanjo, et al., 2011). All these studies indicate that cassava needs irrigation at 100% level to realize the maximum tuber yield during summer months.

TABLE 7.1 Tuber Yield and B: C Ratio as Affected by Levels of Micro irrigation in Cassava.

Irrigation levels	Tuber yield (t ha^{-1})	B:C ratio
I_1 100% CPE	43.94[a]	4.27
I_2 80% CPE	38.00[b]	3.56
I_3 60% CPE	31.98[c]	2.83
Surface flood	32.0[c]	3.44
Rainfed crop	11.0[d]	1.14
CD (0.05)	5.39	—

7.3.3 WATER PRODUCTIVITY

On an average, the crop consumed 540 mm of water for the whole season at 100% PE followed by 478 mm at 80% PE and 416 mm at 60% PE. An effective rainfall of 230 mm was received during the period. The data clearly revealed that supplementary irrigation during summer months increases cassava tuber yield, though the crop is reported to be drought tolerant. Increase in water use efficiency (WUE) in cassava with drip irrigation compared to the surface method has been reported by earlier workers (Amanullah, et al., 2006; Odubanjo, et al., 2011).

Water productivity was worked out for all the irrigation levels. Among the irrigation levels, the value was maximum in I_1 (8.2 kg m^{-3}) followed by I_2 and then I_3. Under surface irrigation, the water productivity was minimum (2.6 kg m^{-3}), because a lot of water was wasted through runoff and percolation and was not reflected in tuber yield, which resulted in a very low water productivity (Table 7.2). This shows the advantages of micro irrigation in the active root zone as reported in many crops. Moreover, micro irrigation recorded water saving of 69, 75 and 81% under I_1, I_2 and I_3 irrigation levels, compared to surface flood method.

TABLE 7.2   Water Productivity and Water Saving Estimations under Different Irrigation Levels in Cassava

Item	I_1	I_2	I_3	Rainfed	Surface flood
Irrigation water applied (mm)	310	248	186	—	1000
Effective rainfall (mm)	230	230	230	230	230
Total water used (mm)	540	478	416	230	1230
Tuber yield (t ha^{-1})	44.00	38.00	32.00	11.00	32
Water productivity (kgm^{-3})	8.2	7.9	7.6	4.2	2.6
Water saving (%)	69	75	81	-	-

I_1: 100% CPE, I_2: 80% CPE, I_3: 60% CPE

7.3.4 WATER REQUIREMENT

Pan evaporation ranged from 4–5 mm per day during summer months in Kerala. Water requirement of cassava was based on the level of productivity at different levels of water used based on pan evaporation values. The data revealed that a short duration variety during summer season requires an average of 3mm water per day for achieving maximum productivity under humid tropical conditions of Kerala. However, water requirement varies with the stage of the crop and the growth pattern. The requirement ranges from 1.3 mm during initial stages of sprouting to canopy develop-ment (one MAP), then 3.4 mm during the peak period coinciding with maximum growth and tuber development stage (2–5 MAP) and then again coming down to a lower level of 1.3 mm towards senescence and maturity (6 MAP). Irrigation may be stopped before harvesting to hasten maturity of tubers.

7.3.5   ECONOMICS

As in yield performance, benefit: cost ratio followed a declining trend with decreasing levels of irrigation (Table 7.1). Though initial investments under micro irrigation are high compared to surface irrigation, yet the enhanced tuber yield, water saving, and higher water productivity realized under micro irrigation justify this investment.

7.4   SUMMARY

Reduced availability of irrigation water is a major constraint for tuber crops production due to more demand. This assures increasing the water use efficiency of tuber crops through water saving techniques by judicious and timely use of irrigation water. The findings in this chapter revealed that cassava responds well to supplementary irrigation, especially during summer months. Drip irrigation at the rate 100% of pan evaporation resulted in a four-fold increase in tuber yield of cassava compared to no irrigation. Though established to be a drought-tolerant crop, a short dura-tion variety raised during summer months with micro irrigation resulted in almost 30% increase in tuber yield and 50% saving in water consumption compared to surface flood irrigation.

KEYWORDS

- cassava
- micro irrigation
- tuber yield
- water productivity

REFERENCES

Allen, R., & Pruitt, W., (1991). FAO 24 Reference evapotranspiration factors. *J. Irrig. Drain Eng.*, *117*, 758–773.

Amanullah, M. M., Mohamed-Yassin, M., Vaiyapuri, K., Somasundaram, E., Sathyamoorthi, K., & Padmanathan, P. K., (2006). Growth and yield of cassava as influenced by drip irrigation and organic manures, *Research Journal of Agriculture and Biological Sciences, 2*, 554–558.

Edoga, R. N., & Edoga, M. O., (2006). Design of drip irrigation set for small vegetable gardens. *Notulae Botanicae Horti Agrobotanici Cluj-Napoca, 34*, 1842–1859.

Khatib, S., Safaa A., Mansour, A., & Sahar, A. S., (2007). Performance of cassava plant in sandy soil under different irrigation intervals, water quantities and plant spacing. *J. Agric. Sci. Mansoura Univ., 32*, 9207–9220.

Mogaji-Kehinde, O., Olotu, Y., Oloruntade, A. J., & Afuye, G. G., (2011). Effect of supplemental irrigation on growth, development and yield of cassava under drip irrigation system in Akure, Oondo State, NIgeria. *Journal of Sciences and Multidisciplinary Research, 3*, 62–73.

Nayar, T. V. R., Kabeerathumma, S., Potty, V. P., & Mohankumar, C. R., (1993). Recent progress in cassava Agronomy in India. In: *Proc. 4thRegional Workshop on Cassava Breeding Agronomy Research and Tech. Transfer in Asia*, Trivandrum, Kerala, India, pp. 61–83.

Odubanjo, O. O., Olufayo, A. A., & Oguntunde, P. G., (2011). Water use, growth and yield of drip irrigated cassava in a humid tropical environment. *Soil & Water Res., 6*, 10–20.

Ramanujam, T., (1994). Water management in cassava. In: Chadha, K. L., & Nayar, G. G., (eds.), *Advances in Horticulture: Vol. 8 – Tuber Crops* (pp. 333–342.). Malhotra Pub. House, New Delhi, India

SAS, (2010). SAS Institute Inc., Cary, NC, USA.

CHAPTER 8

WATER USE EFFICIENCY OF PEARL MILLET AND NAPIER HYBRID GRASS UNDER IRRIGATION REGIMES AND NITROGEN FERTIGATION

S. ALAGUDURAI

ABSTRACT

This chapter reveals that paired row system of planting (60 x 50; 90 x 50 cm), 150% pan evaporation along with 125% N fertigation recorded higher green fodder yield of Bajra (pearl millet (*Pennisetum glaucum*)) and Napier hybrid grass (*Pennisetum purpureum*) var. CO(CN)4. Drip irrigation resulted in the considerable saving of irrigation water beside enhancing the water use efficiency over the surface method of irrigation.

8.1 INTRODUCTION

Livestock plays an important role by providing employment and supplementary family income and it provides balanced and nutritional food in the form of milk, meat, and egg to millions of people. India accounts for 15% of the world's livestock population in 2% of the world's geographical area indicating huge pressure on land (Ghosh, 2014). Although India has a very large population of livestock, yet the productivity of milk and other livestock products per animal is relatively low compared to other countries around the world. One of the main reasons for this low productivity of our livestock is either malnutrition or under-nutrition or both, besides the inherent genetic potential of the animals.

Feed and fodder are the key factors for enhancing milk production and yield level. Profitable livestock farming depends mainly on the adequate availability of fodder throughout the year. The technology of growing year-round fodder production has helped the livestock farmers to sustain milk and meat production with the judicious and economical use of concentrates and a considerable reduction in the cost of production. The economically competitive and productive yield potential of crossbred milch animals can be exploited through feeding nutritious green fodder (Velayudham, et al., 2011). Bajra and Napier hybrid grass are potential perennial sources of green fodder. Both are popular owing to high yield, palatability, and adaptability to varying soil and climatic conditions (Faruqui, et al., 2009). Bajra and Napier hybrid grass CO(CN)4 are found to tiller profusely and yield more than other varieties.

In Tamil Nadu, the land area utilized for growing fodder is very negligible, accounting only 1.6% of the total cultivated area (Season and Crop Report, 2012). There is a great need to maintain the regular well-balanced supply of more nutritious feed and fodder for stall feeding milch animals, which would accelerate the growth of milk production in the state.

Nitrogen is the most important plant nutrient required for crop production and is required in large quantities (Balasubramanian et al., 2010). Bajra and Napier hybrid grass are heavy nutrient feeder and due to their multi-cut ability, the nutritional requirement is also very high as it gives a better response to fertilizer application, especially N (Pathan, et al., 2012).

In Tamil Nadu, Bajra, and Napier hybrid grass are mainly grown under irrigation. Fodder production in the irrigated area is characterized by water deficit, especially in summer. Thus, the efficient use of available water through drip irrigation is successful in augmenting fodder production. By introducing drip irrigation, it is possible to increase the yield potential of crops by three times with the same quantity of water, by saving about 45 to 50% of irrigation water and increasing the productivity by about 40 percent.

Information on the optimum crop geometry to explore the available resources, water requirement and N fertigation schedule for Cumbu Napier hybrid grass is meager. Therefore, the research study in this chapter has contemplated to practice intensive and year-round fodder production through paired row drip fertigation method to solve the problem of water scarcity, enhancing water use efficiency and fodder productivity.

8.2   MATERIALS AND METHODS

The field experiment was carried out at Krishi Vigyan Kendra farm, Tamil Nadu Veterinary College and Research Institute Campus (TANUVAS), Namakkal, Tamil Nadu during 2012–13 and 2013–14. The soil at the experimental site is sandy loam having 18.76% field capacity, 7.74% permanent wilting point and bulk density of 1.04 (Mg m^{-3}). The soil has a pH value of 7.96, organic carbon content of 0.27% and EC of 0.19 dSm^{-1}. The soil is low in available nitrogen (101 kg ha^{-1}), medium in available phosphorus (15.18 kg ha^{-1}) and available potassium (215 kg ha^{-1}). The experiment was laid out in a randomized block design with three replications. Treatments comprised of two methods of planting (paired row drip system (60/90 ×50 cm and 60/120 ×50 cm)), two levels of irrigation regime (Drip irrigation at 100% PE and 150% PE) and three levels of N fertigation (75, 100 and 125% of recommended dose of N fertilizer). Treatments details were:

T_1	Paired row drip system (60/90 ×50 cm) + Drip at 100% PE + N fertigation at 75% RDF
T_2	Paired row drip system (60/90 ×50 cm) + Drip at 100% PE + N fertigation at 100% RDF
T_3	Paired row drip system (60/90×50 cm) + Drip at 100% PE + N fertigation at 125% RDF
T_4	Paired row drip system (60/90 ×50 cm) + Drip at 150% PE + N fertigation at 75% RDF
T_5	Paired row drip system (60/90 ×50 cm) + Drip at 150% PE + N fertigation at 100% RDF
T_6	Paired row drip system (60/90×50 cm) + Drip at 150% PE + N fertigation at 125% RDF
T_7	Paired row drip system (60/120×50 cm) + Drip at 100% PE + N fertigation at 75% RDF
T_8	Paired row drip system (60/120×50 cm) + Drip at 100% PE + N fertigation at 100% RDF
T_9	Paired row drip system (60/120×50 cm) + Drip at 100% PE + N fertigation at 125% RDF
T_{10}	Paired row drip system (60/120×50 cm) + Drip at 150% PE + N fertigation at 75% RDF
T_{11}	Paired row drip system (60/120×50 cm) + Drip at 150% PE + N fertigation at 100% RDF
T_{12}	Paired row drip system (60/120×50 cm) + Drip at 150% PE + N fertigation at 125% RDF
T_{13}	Control: Surface irrigation (5 cm depth) + soil application of N at 100% RDF with a spacing of 60 x 50 cm

The recommended dose of fertilizers (150:50:40 NPK kg/ha and 75 kg of N) was applied after each harvest. Nitrogen alone was applied through drip irrigations at various levels as per the treatments and entire P and K fertilizers were applied as basal.

Two budded stem cuttings of Bajra and Napier hybrid grass CO(CN)4 were used for planting and the planting was done on 27/09/2012. The paired row system of planting was followed as per the treatments schedule. All the cultural practices other than the treatments were followed as per the recommendations of Crop Production Guide, TNAU (Crop Production Guide, 2012). Gap filling was carried out ten days after planting for maintaining uniform plant population.

8.2.1   LAYOUT OF DRIP SYSTEM

Under a paired row planting system, one 16 mm lateral was laid out for each two rows of Bajra and Napier hybrid grass with a lateral spacing of 1.5 m for 60/90 x 50 cm and 1.8 m for 60/120 x 50 cm paired row system with one dripper (4 lph) for two plants at 50 cm apart along the lateral. Drip irrigation was scheduled based on 2 days of cumulative pan evaporation (CPE) and fertigation was carried out once in four days based on nutrient uptake pattern at different phonological growth phases of the crop.

For surface irrigation treatment, the recommended dose of fertilizers 150:50:40 kg NPK ha^{-1} was applied. At the time of planting, 50% N and entire dose of P and K fertilizers were applied as basal and the remaining 50% N was top dressed at 25 days after planting for the main crop and 75 kg of N was applied immediately after each cutting for ratoon crop. Under surface irrigation treatments, irrigation was given before planting followed by life irrigation on the third day. Scheduling of subsequent irrigation was based on IW/CPE ratio of 0.80 (i.e., cumulative pan evaporation value reached 62.5 mm during throughout crop period). Daily pan evaporation rate was recorded from the standard USWB Class A Open pan evaporimeter that was installed in the field itself. Irrigation was given to the depth of 5 cm.

For drip irrigation treatments, initial soak – irrigation was given uniformly before planting for good germination and subsequent irrigations were scheduled once in two days and applied each time as per the treatment schedule. For drip fertigation system, the operating pressure was maintained at 1.0 kg cm^{-2}.

The first cutting was done 75 days after planting and subsequent cutting was carried out at every 45 days interval. In total 14 harvests were taken from two consecutive years. Five clumps were selected at random from each plot and were tagged for recording the biometric and yield parameters. The data on the yield parameters from different cuts were pooled and subjected to statistical analysis.

8.3 RESULTS AND DISCUSSION

The data on green fodder yield of Bajra and Napier hybrid grass are presented in Table 8.1. Paired row planting system, drip irrigation regime and N fertigation levels had significantly influenced the green fodder yield compared to the surface method of irrigation along with the conventional method of fertilizer application.

TABLE 8.1 Effect of Crop Geometry, Irrigation Regimes and N Fertigation Levels on Green Fodder Yield of Bajra and Napier Hybrid – CO(CN)4 During 2012–13 and 2013–14 (Pooled Analysis)

Treatments	Green fodder yield (tons ha^{-1})		Treatments	Green fodder yield (tons ha^{-1})	
	2012–13	2013–14		2012–13	2013–14
T$_1$	326.32	305.15	T8	321.91	309.80
T$_2$	346.92	327.36	T9	341.35	326.17
T$_3$	368.17	349.83	T10	312.17	296.56
T$_4$	332.03	312.32	T11	330.35	316.59
T$_5$	356.50	338.52	T12	349.53	330.99
T$_6$	388.06	366.35	T13	280.94	252.55
T$_7$	302.98	288.48			
Mean	**335.15**	**316.97**			
S.Ed.	**14.134**	**10.89**			
C.D. (P = 5%)	**29.71**	**22.47**			

Paired row system of planting (60/90 x 50 cm) with drip irrigation at 150% PE and N fertigation at 125% RDN (T$_6$) registered significantly higher green fodder yield of 388.1and 366.3 t ha^{-1} during 2012–13 and 2013–14, respectively, and it was comparable with paired row system of planting (60/90 x 50 cm) with drip irrigation at 100% PE and N fertigation at 125% RDN (T_3), which recorded 366.2 and 349.8 t ha^{-1} during 2012–13 and 2013–14, respectively, and it was 35.4 and 40.4% higher when compared to surface method of irrigation with 100% RDF through conventional method of fertilizer application (280.9 and 252.5 t ha^{-1} during 2012–13 and 2013–14, respectively).

This may be mainly attributed to increased nutrient availability and subsequent absorption by the crop under the higher moisture condition coupled with frequent nutrient delivery through fertigation and consequent better formation and translocation of assimilates from source to sink might have increased growth and yield parameters (*viz.*, plant height, number of tillers m^{-2}, number of leaves m^{-2} and leaf stem ratio and the beneficial effects of nitrogen on cell division and elongation, formation of nucleotides and co-enzymes), which resulted in increased meristematic activity and photosynthetic area and hence more production and accumulation of photosynthates, yielding higher green fodder and dry matter yield of Bajra and Napier hybrid grass. These results are in conformity with the findings of Malarvizhi and Fazlullahkhan (2000), Ayub et al. (2009) and Vinayraj (2013) in hybrid Napier grass.

Increasing amounts of irrigation water from 100 to 125%, etc., increased the fresh biomass yield by 14% (from 19 to 22 t ha^{-1}) in forage maize was observed by Al-Dhuhll et al., (2013). Hassan et al. (2010) reported the application of 140 kg N ha^{-1} through fertigation recorded higher green and dry fodder yield than other N fertigation levels of 100 and 60 kg ha^{-1} in fodder maize.

Lower green fodder yield was registered under surface irrigation with soil application of 100% recommended dose of NPK. This might be attributed to a decrease in synthesis of metabolites and reduction in absorption and translocation of nutrients from the soil to plant due to unfavorable moisture content. The physiological response of plants by decreased cell division and cell elongation under moderate moisture stress at wider irrigation intervals might have also contributed to reduced green fodder yield under surface method of irrigation. The results are in conformity with the findings of Wood and Finger (2006).

The data on water use efficiency (WUE) of different treatments are shown in Table 8.2. WUE varied due to different crop geometry, drip irrigation regimes and N fertigation levels. Optimum irrigation regimes (100% PE) with wider paired row planting (60/120 x 50 cm) had a favorable and marked influence on WUE. Drip irrigation method favorably influenced the WUE compared to surface irrigation. Among the different treatments, paired row planting at 60/120 x 50 cm with drip irrigation at 100% PE and N fertigation at 125% RDN (T_9) registered significantly higher WUE of 311.7 and 402.2 kg-mm ha^{-1} during 2012–13 and 2013–14, respectively, and followed by paired row system of planting (60/90 x 50 cm) with drip irrigation at 100% PE and N fertigation at 125% RDN (T_3), which recorded 297.4 kg-mm ha^{-1} during 2012–13 and paired row planting at 60/120 x 50 cm with drip irrigation at 100% PE and N fertigation at 100% RDN (T_8), which recorded 382.0 kg-mm ha^{-1} during 2013–14. Significantly lower WUE of 150.5 and 172.4 kg-mm ha^{-1} was observed under surface method of irrigation with 100% RDF through the conventional method of fertilizer application (T_{13}).

TABLE 8.2 Effect of Crop Geometry, Irrigation Regimes and N Fertigation Levels on Water Use Efficiency of Bajra and Napier Hybrid Grass During 2012–13 and 2013–14

Treatments	WUE (kg/ha-mm^{-1})		Treatments	WUE (kg-mm/ha)	
	2012–13	2013–14		2012–13	2013–14
T_1	263.6	328.3	T8	293.9	382.0
T_2	280.2	352.2	T9	311.7	402.2
T_3	297.4	376.4	T10	214.9	267.9
T_4	199.2	241.3	T11	227.4	286.0
T_5	213.9	263.5	T12	240.6	299.0
T_6	232.8	285.2	T13	150.5	172.4
T_7	276.6	355.7			

WUE can be increased either by increasing the yield or reducing the quantity of water applied. Increase in water use efficiency in drip system over furrow irrigation was mainly due to the controlled water release near the crop root zone leads to considerable saving of irrigation water, a greater increase in yield of crops and higher nutrient use efficiency. This agrees with the findings of Bobade (1999).

Among the irrigation regimes, 100% PE registered higher WUE than 150% PE. Water use efficiency was decreased with increases in irrigation regime. This agrees with the findings of Al-Suhaibani (2006) in Sudan grass, Simsek et al. (2011) in fodder maize, Idris et al. (2013) in alfalfa fodder production. Higher consumption of water with lesser green fodder yield resulted in lower WUE under surface irrigation with soil application of recommended dose of fertilizer.

8.4 SUMMARY

The drip irrigation was scheduled once every two days with the computed quantity of water as per the treatment schedule of 100 and 150% PE and fertigation were carried out once every four days based on different phonological growth phases of main and ratoon crop. Nitrogen alone was fertigated through Venturi in the form of urea at various levels as per the treatments and fertigation were stopped 20 days before harvesting of fodder to avoid higher nitrate accumulation and entire dose of P and K fertilizers were applied as basal.

Paired row system of planting at 60/90 x 50 cm with drip irrigation at 150% PE and N fertigation at 125% RDN registered significantly higher green fodder yield (388.1and 366.4 t ha⁻¹) and the percentage yield increase was up to 35.4 and 40.4% over surface method of irrigation with 100% RDF through conventional method of fertilizer application during 2012–13 and 2013–14, respectively.

Drip irrigation method favorably influenced the WUE compared to surface irrigation method. Irrigation regime at 100% PE registered more WUE than 150% PE. Treatment combination of paired row planting at 60/120 x 50 cm with drip irrigation at 100% PE and N fertigation at 125% RDN registered significantly higher WUE during 2012–13 and 2013–14, respectively. Surface method of irrigation with the conventional method of planting and fertilizer application registered a relatively lower WUE during both the years.

KEYWORDS

- Bajra
- drip fertigation
- green fodder
- Napier hybrid grass
- paired row planting
- water use efficiency

REFERENCES

Al-Suhaibani, N. A., (2006). Effect of irrigation intervals and nitrogen fertilizer rates on fresh forage yield of Sudangrass (*Sorghum sudanense* (Piper) Stapf.). *Res. Bult., 142*, 5–17.

Al-Dhuhll, H., Schmitz, G. H., Linnartz, F., Schutze, N., Grundmann, J., Kloss, S., & Pistorius, M., (2013). Optimal irrigation scheduling for fodder crops under multiple resource constraints in an arid zone. *Environment Proceeding on ICWRER*, pp. 16–23.

Ayub, M., Athar, M., Nadeem, A., Tahir, M., Ibrahim, M., & Aslam, M. N., (2009). Effect of nitrogen application and harvesting intervals on forage yield and quality of pearl millet (*Pennisetumamericanum* L). *Pak. J. Life Soc. Sci., 7*(2), 185–189.

Balasubramanian, V., Raghauram, N., Abrol, I. P., Sachdev, M. S., Pathak, H., & Singh, B., (2010). *Reactive Nitrogen: Good, Bad and Ugly*. Comprehensive Status Report, SCON-ING, New Delhi, p. 55.

Bobade, S. V., (1999). *Studies on Drip Irrigation and Nitrogen Levels on the Water Use and Yield of Eggplant (Var. CO–2)*. MSc. (Ag.) Thesis submitted to Tamil Nadu Agricultural University, Coimbatore, p. 139.

Crop Production Guide, (2012). Tamil Nadu Agriculture University and Directorate of Agriculture, Chennai, p. 300.

Faruqui, S. A., Sunilkumar, T., & Singh, D. N., (2009). Napier Bajrahybrid. In: *Excellent Perennial Fodder*, AICRP on FC, IGFRI, Jhansi, p. 16.

Ghosh, P. K., (2014). Forage resource development in India: Strategies and issues. In: *Proc. in Indian Society of Agronomy "Agricultural Diversification for Sustainable and Environmental Security"*, PAU, Ludhiana, Punjab, India, p. 43–46.

Hassan, S. W., Oad, F. C., Tuni, S. D., Gandhi, A. W., Siddiqui, M. H., & Jagirani, A. W., (2010). Impact of nitrogen levels and application methods on agronomic, physiological and nutrient uptake traits of maize fodder. *Pak. J. Bot., 42*(6), 4095–4101.

Idris, A. Y., El-Nadi, A. H., Dagash, Y. M. I., & Ali, S. A. M., (2013). Comparative study of Lucerne (*Medicagosativa* L.) under drip and sprinkler Irrigation. *Universal Journal of Agricultural Research, 1*(2), 17–23.

Malarvizhi, P., & Fazlullah, K. A. K., (2000). Effect of graded dose of nitrogen and varying cutting intervals for higher biomass production in Bajra Napier hybrid grass CO(3). In: *National Seminar on Strategy for Maximization of Forage Production*, BCKV, West Bengal, pp. 12–16.

Pathan, S. H., Tumbare, A. D., & Kamle, A. B., (2012). Impact of planting material and cutting management and fertilizer level on nutritional quality of Bajra Napier hybrid. *Forage Res., 38*(2), 74–79.

Season and Crop Report, (2012). *Department of Economics and Statistics*, Chennai, pp. 352.

Simsek, M., Abdullah, C., Denek, N., & Tonkaz, T., (2011). The effects of different irrigation regimes on yield and silage quality of corn under semi-arid conditions. *African J. Biotech., 10*(31), 5869–5877.

Velayudham, K., Babu, C., Iyanar, K., & Kalamani, A., (2011). Impact of plant geometry and fertilizer levels on Bajra Napier hybrid grass. *Indian J. Agric. Sci., 81*(6), 575–577.

Vinayraj, D. J., & Palled, Y. B., (2013). Response of hybrid Napier genotypes to nitrogen levels. *Karnataka J. Agric. Sci., 27*(1), 74–75.

Wood, M. L., & Finger, L., (2006). Subsurface drip irrigation for alfalfa fodder production. *Australian Journal of Experimental Agriculture, 46*(12), 1605–1614.

PART III

Performance of Fertigated Rice under Micro irrigation

CHAPTER 9

STATUS OF DRIP IRRIGATION RESEARCH IN RICE: A REVIEW

R. NAGESWARI, R. CHANDRASEKARAN, and T. SARANRAJ

ABSTRACT

Rice cultivation consumes more than 75% of the water used in agriculture. Around 98% of rice fields around the Globe are in small farms. In India, rice is cultivated in 44 million ha with an average yield of 2 tons/ha. Rice is mainly irrigated by flooding. Irrigated rice uses an estimated 34 to 43% of the total World's irrigation water, or about 24 to 30% of the entire World's developed freshwater resources. It leads to generate the greenhouse gases, methane, and nitrous oxide and leaches down the nitrogen nutrient. Many researchers in the southern Tamil Nadu found that about 2.9 mg of nitrous oxides are generated per m^2 daily. Whereas, the paddy crop irrigated through drip irrigation produce a lesser amount of about 0.5 mg of nitrous oxides per m^2 daily. In the USA, the Ben Gurion University conducted an experiment on drip irrigation in rice at the Gonaway Ranch for the first time. A study conducted at the University of Agricultural Sciences, Bangalore reported that consistent use of high water potency (91.01 kg/ha.cm^{-1}) for drip fertigated rice with 100% recommended a dose of fertilizers. At Tamil Nadu Agricultural University, research trials conducted by the Water Technology Centre indicated that around 50% of the water use could be minimized by drip irrigation in paddy. Similar studies were also conducted at TRRI, Aduthurai, and ADAC&RI, Trichy to test the feasibility of growing rice under drip irrigation. The demonstration conducted on drip irrigation in the Amaravathy sub-basin under TN-IAMWARM project enabled the farmer, Mr. M. Parthasarathy, to get 'Innovative Rice Farmer Award for the year 2015' by the Indian Institute of Rice Research, Hyderabad. All these research studies conclude that 20% more available land can be used for growing rice Worldwide with the advent of drip irrigation, using its ability to grow in various soil types and changing topography besides water saving of 30–40%.

9.1   INTRODUCTION

Rice is grown in 115 countries and consumes more than 75% of the water used in agriculture. About 98% of rice fields are occupied by small farmers. In India, rice is cultivated in 43.2 million ha with a production of 157.2 million tons and an average yield of 2.0 tons/ha (FAO, 2015). Under this, irrigated transplanted rice is grown in 25.1 m-ha (57%), dry seeded rice in 12.26 m-ha, and irrigated direct sown rice in 6.0 m-ha. Currently, India's foodgrain output is at 247 million tons. By 2050, it will have to produce 494 million tons. The water withdrawn for agriculture in India is 90.41% of total water used today and is expected to shrink to 71.6% in 2025 and to 64.6% in 2050 (FAO, 2015). That is unlikely to happen unless farmers adopt better farming techniques. Irrigation alone contributes to 60 to 100% of grain yields and this is the only real option to enhance foodgrain output. Drip irrigation in the recent era of modernization of agriculture is an innovative, reliable, sustainable irrigation technology with a more effective use of resources, leading to higher overall yield production.

This chapter focuses on the literature review on drip irrigation technology in rice.

9.2   ISSUES OF FLOOD IRRIGATION IN RICE PRODUCTION

Rice has been mainly irrigated for 5,000 years using flood irrigation. Irrigated rice uses estimated 34 to 43% of the total World's irrigation water or about 24 to 30% of the entire World's developed freshwater resources. Almost 90% of the freshwater consumed in India is in agriculture. The bulk portion of about 70% is used to cultivate paddy. The flooding method is a major reason for the country earning the dubious distinction of being the second largest producer of methane in the world, after China, according to the Global Methane Initiative (GMI) that aims to reduce global methane emissions. The water productivity of paddy in India is the lowest in the world, at 150 g of paddy per 1,000 liters, resulting in an average output of 2.1 tons per ha (FAO, 2015).

The dominant greenhouse gases generated in rice growth are methane and nitrous oxide. Methanogens produce methane gas under anaerobic conditions in the rice fields, while nitrifying and denitrifying bacteria that operate under anoxic conditions generate nitrous oxides. In rice fields irrigated by flooding, research works in the southern Tamil Nadu State found that about 2.9 mg of nitrous oxides are generated per m²daily. Whereas, those fields irrigated by drip methods only produce about 0.5 mg of nitrous oxides per m²daily. The total annual CH_4 emission (both from natural and anthropogenic terrestrial sources to the atmosphere) is about 500 Tg/year (CH_4) (1 Tg = 10 million tons) (TNAU, 2013). The contribution of natural and man-made wetlands to this global total varies between 20 and 40%. Rice agriculture accounts for some 17%of the anthropogenic CH_4 emissions according to Institute of Biogeochemistry and Pollutant Dynamics, Zurich. The roots absorb hazardous metals from the soil in the presence of flood water.

Rice plants under aerobic systems undergo several cycles of wetting and drying conditions (Matsuo et al., 2009). Such a mild plant water stress at vegetative growth stage can decrease the number of tillers (Cruz et al., 1986). There were significant differences in rooting characteristics, especially deep rooting depth, and root biomass, among various (aerobic and upland) rice varieties (Kondo et al., 2003). There are only a few attempts to address the physiological responses of rice and critical analysis of various yield components to aerobic and drip irrigated condition (Belder et al., 2004; Bouman et al., 2005). Poor root systems and root function may limit water absorption and decrease leaf water potential (Pandian, 2013) under aerobic cultivation. In the current scenario, drip irrigation offers a viable and alternate water-saving system for rice.

9.3   ADVANTAGES OF DRIP IRRIGATION IN RICE

- Conserving energy use for pumping up to 52%.
- Good aeration in the soil which decreases the accumulation of heavy metals in the grain.
- Higher fertilizer use efficiency.
- Higher water use efficiency.
- higher yield of straw fodder.
- Irrigation water saving up to 40% about to 10,000 to 20,000 m³/ha and saves 30% of nutrients.

- Low incidence of pests and diseases.
- Minimum requirement of land preparation and avoidance of soil compaction.
- Possibility of growing rice in saline water.
- Possibility to grow rice under various soil types and changing topography which leads to add, potentially, 20% more available land for growing rice worldwide.
- Prevention of Methane emission and the protection of the environment in the absence of standing water.
- Protection of the environment from pollution from leached and washed Nitrates and significantly reduced greenhouse gas emissions and groundwater pollution.
- Yield enhancement up to 20 to 50% due to uniform water and nitrogen distribution.

9.4   PROGRESS OF DRIP IRRIGATION RESEARCH IN RICE

Established in 1965, drip irrigation originated with the discovery of water engineer Simcha Blass. In 2006, Netafim approached the aerobic rice cultivation with fertigation by drip. A similar study about the relationship of traditional rice field cultivation and greenhouse gas emissions occurred in Thailand, at the King Mogkut's University of Technology in Thonburi in 2005. In that study, researchers looked at various options for reducing the enormous greenhouse gas emissions emanating from the paddy fields, such as draining the flooded fields and altering fertilizer application methods. The American study, presented to the Pew Center on Global Climate Change in 2006 by researchers from Colorado State University, Montana State University, and the National Renewable Energy Laboratory, explored the production of greenhouse gases in the US agricultural sector. As in the other studies, these researchers also concluded that a more efficient use of nitrogen can reduce nutrient runoff and improve water quality in both the ground and surfaces.

The first aerobic rice by drip fertigation in Italy started in 2010. In the USA, the Ben Gurion University conducted an experiment on drip irrigation in rice at the Gonaway Ranch for the first time. For the first time, Jain Irrigation systems experimented the technology for paddy cultivation in India and abroad during 2010. A collaborative research program on drip irrigation in paddy is being carried out at Tamil Nadu Rice Research

Institute, Aduthurai, and Anbil Dharmalingam Agricultural College and Research Institute, Trichy to test the feasibility of growing rice under drip irrigation with the financial support from Netafim Irrigation India Pvt Ltd. Similar studies were also conducted in Australia, Brazil, China, Italy, Japan, Spain, Taiwan, Thailand, Turkey, Ukraine, and USA (Texas).

9.5 STATUS OF DRIP IRRIGATION IN RICE CULTIVATION

The experiment conducted at Xinjiang province, China demonstrated that the plastic mulching with drip irrigation (DI) has greater water saving capacity and lower yield and economic benefit gaps than the non-flooded irrigation incorporating plastic mulching with furrow irrigation (FIM) and non-mulching with furrow irrigation (FIN) treatments compared with the Conventional Flooding (CF) and would therefore be a better water-saving technology in areas of water scarcity (He et al., 2013). The DI treatment had a higher grain yield and harvest index, more effective tillers, more roots in topsoil, higher WUE and greater economic benefit compared with the FIM and FIN treatments. There are three groups of water saving technologies followed in rice cultivation:

- The first group includes the continuously saturated soil cultivation system, the rice intensification system (Uphoff et al., 2002); and the alternate wetting and drying system (Liang et al., 2003). These cultivation systems retain high soil water contents, or in some growth stages, flooded soils, therefore water losses are high (Peng et al., 2006).
- The second group is known as "aerobic rice," in which rice, like upland crops, is grown under non-flooded conditions with adequate inputs and supplementary irrigation when rainfall is insufficient (Rekha et al., 2015; Zhang et al., 2008). Because of the significant reduction in seepage, percolation, and evaporation, this technology allows for greater WUE and high water saving compared with traditional flooded irrigation (Bouman et al., 2002).
- The third group is ground cover rice production systems (GCRPSs) (Tao et al., 2006), which are basically "aerobic rice" systems. They utilize plastic mulching or straw mulching in the cultivation system. Under these ground cover conditions, evaporation can be effectively reduced compared with bare land conditions, therefore GCRPSs have higher WUE than the "aerobic rice" (Quin et al., 2006).

It was suggested that plastic mulching cultivation has great potential to substantially save water resources at a high grain yield level compared with traditional flooded irrigation because of the warming and water retention effects of plastic mulching. Therefore, when the stressful factors are water deficits and/ or low soil temperature during the vegetative growth stage, plastic mulching cultivation could be a promising technology to promote rice grain yield formation and WUE (Liang et al., 2003).

Under non-flooded irrigation, the root-zone environment changes from being anaerobic to aerobic. Compared with traditional flooding, fewer roots are distributed in the topsoil layer, while more roots tend to be distributed in deeper soil layers (Kato et al., 2011). It is also widely believed that the root distribution zone moves upward under drip irrigation when compared with furrow irrigation (Hodson et al., 1990).

Using drip irrigation, with 5 to 6 million liters of water per ha, around 5 to 6.5 tons of rice could be produced. This was more effective than the existing methods. Under the 'aerobic rice' method, where rice was cultivated in the garden land, around 6 to 6.5 million liters of water were used per hectare to produce 4.5 to 5.5 tons of rice. The popular System of Rice Intensification (SRI) method used around 12 to 15 million liters per hectare to produce 7 tons of rice (TNAU, 2013; Pandian, 2013).

In SRI, drip irrigation gave higher field water use efficiency of 0.45 kg grain/m^3 of water, which was higher than that of traditional method (0.30 kg grain/m^3 of water) and DSR method with drip irrigation (0.27 kg grain/m^3 of water) at Indian Agricultural Research Institute (IARI), New Delhi (Basavaraj, 2013). Whereas, irrigation use efficiency was higher in SRI method with drip irrigation (0.67 kg grain/m^3 of water). Considering the conservative amount of fertilizer application, less amount of fertilization in a normal paddy field, the yield potential of rice could be improved by increasing the amount of fertilizer as a top application in drip irrigation system (Adekoya et al., 2014).

Application of 100% RDF through drip fertigation with water-soluble fertilizer recorded higher growth parameters (viz., plant height (56.70 cm), number of tillers hill^{-1} (50.43), number of leaves hill^{-1} (191.43), total dry matter production (138.39 g hill–1)) and grain and straw yield of 6503 and 9285 kg ha^{-1}, respectively (Rekha et al., 2015; Vijaykumar, 2009). Drip irrigation technique favored the growth of the paddy during the non-monsoon period under sodic soil condition of Manikandam block in Trichy district (Pandiyarajan, 2016).

A field experiment conducted at Zonal Agricultural Research Station, Bangalore revealed that application of 25% N & K from sowing to 30 DAS + 25% from 31 to 50 DAS + 25% from 51 to 80 DAS + 25% from 81 to 105 DAS through drip irrigation recorded significantly higher grain yield (11.0 Mg ha^{-1}) over-application of N & K as per package of practice (9.2 Mg ha^{-1}), which was 20.2% higher over-application of N & K as per UAS (B) recommendations through drip irrigation and 70.1% over soil application (Prabhudeva et al., 2016). Studies conducted at Tamil Nadu Agricultural University, Coimbatore indicated that the lateral spacing of 0.8 m with 1.0 lph drippers is best for rice cultivation in enhancing the growth, physiology, grain yield and water productivity (Theivasigamani, 2017).

9.6   SUCCESS STORIES UNDER DRIP IRRIGATION: CASE STUDIES

Rajesh Vijay, a native of Bhadana village, 15 km from Kota town of Rajasthan used drip irrigation method of cultivating rice with the help of M/S. Jain irrigation systems to double the area under rice with the same quantity of water beside saving in water and electricity consumption to the tune of 40% each. The rice yield has increased by 25%, earning Rs. 6,000 more per acre per crop cycle (Rs. 60.00 = one US$) (http://www.businesstoday.in/magazine/cover-story/drip-irrigation-of-paddy-improves-yields-saves-water/story/19084).

The concept of drip fertigation for rice was selected as 'Best Management Practice (BMP)' by Centre for World Solidarity in 2014 (IIRR, 2015). The demonstration conducted on drip irrigation in the Amaravathy sub-basin under the TN-IAMWARM project implemented by the Tamil Nadu Agricultural University enabled the farmer, Mr. M. Parthasarathy, to get 'Innovative Rice Farmer Award for the year 2015' from the Indian Institute of Rice Research, Hyderabad.

The Chinmaya Farmers Club members formed part of a Project conducted by Anna University, Chennai that provided 100% subsidy for procurement and installation of Drip Irrigation Systems and 100% subsidy in the second year for Quality Certified seeds. Vegetables and other cereals farming got higher yield under Drip Irrigation (Amrita, 2014). Mr. V. Annamalai from Pallakollai village of Tiruvannamalai district was able to save 0.9 million liters of water by cultivating paddy under drip irrigation when compared to flooding method (The Hindu, 2013). The success of these few progressive farmers across the country achieved by adopting drip irrigation holds hope for India's food security.

9.7 LIMITATIONS IN THE ADOPTION OF DRIP IRRIGATION IN RICE

Under an aerobic environment in drip-irrigated paddy fields, the presence of more weeds is inevitable. The government does not give a subsidy for drip irrigation for growing paddy. There is a big issue among farmers, who do not come forward to adopt this technology in paddy as they find it difficult to imagine growing this crop without standing water. The wetland ecosystem, which nests more number of birds, fishes, etc., will be disturbed.

9.8 FUTURE RESEARCH NEEDS

- Field level demonstrations, Capacity building, and training should be given a very high priority.
- Identification of selective herbicides with high efficiency suitable for the aerobic environment.
- Research on environmental issues related to the usage of drip materials.
- Research on the influence of drip irrigation on quality of rice.
- Rice varieties of high WUE (water use efficiency) should be identified.

9.9 SUMMARY

Adoption of drip irrigation in paddy is a big step forward in addressing water and food problems. Drip irrigation leads to reduce water use, fertilizer use, power use, manpower, and labor. It can be used in various soil types and topography. It results in the reduction of diseases and pests. The current research study emphasizes the estimation of water use, productivity, and environmental issues. However, more research focus must be given on physiological aspects of rice that are altered by the differential soil environment under drip irrigation to maintain or increase the productivity of rice. Evaluation of cost-effectiveness for usage of drip system for irrigating rice must be made.

KEYWORDS

- **drip irrigation**
- **nutrient use efficiency**
- **paddy**
- **water use efficiency**

REFERENCES

Adekoya, M. A., Liu, Z., Eli, V., Zhou, L., Kong, D., Qin, J., & Luo, L., (2014). Agronomic and ecological evaluation on growing water-saving and drought-resistant rice (*Oryza sativa* L.) through drip irrigation. *Journal of Agricultural Science, 6*(5), 110–118.

Amrita, P., (2014). Chinmaya Organization for Rural Development (CORD). In: *Unpublished Annual Report of Activities*, p. 81.

Basavaraj, R. B., (2013). *Response of Rice Varieties to Planting Methods Under Drip Irrigation*. IARI Publications Series/Report No. T–8848, Water Technology Centre, Indian Agricultural Research Institute, New Delhi, p. 25.

Belder, P., Bouman, B. A. M., & Cabangon, R., (2004). Effect of water saving irrigation on rice yield and water use in typical lowland conditions in Asia. *Agr. Water Manage., 65*, 193–210.

Bouman, B. A. M., Peng, S., Castaneda, A. R., & Visperas, R. M., (2005). Yield and water use of irrigated tropical aerobic rice systems. *Agriculture Water Management, 74*, 87–105.

Bouman, B. A. M., Yang, X. G., & Wang, H. Q., (2002). *Aerobic Rice (Han Dao): New Way of Growing Rice in Water-Short Areas*. 12th ISCO Conference, Beijing, China, p. 8.

Cruz, R. T., O'Toole, J. C., Dingkuhn, M., Yambao, E. B., Thangaraj, M., & De Datta, S. K., (1986). Shoot and root responses to water deficits in rainfed lowland rice. *Australian Journal of Plant Physiology, 13*, 567–575.

FAO, (2018). Aquastat, 2015, http://www.fao.org/nr/water/aquastat/data/query/results. html, Accessed on July 17.

He, H., Ma, F., & Yang, R., (2013). Rice performance and water use efficiency under plastic mulching with drip irrigation. *PLoS One, 8*(12), e83103, https://doi.org/10.1371/journal.pone.0083103 Accessed on August 31, 2018.

Hodgson, A. S., Constable, G. A., & Duddy, G. R., (1990). A comparison of drip and furrow irrigated cotton on a cracking clay soil: Water use efficiency, waterlogging, root distribution and soil structure. *Irrigat. Sci., 11*, 143–148.

IIRR, (2018). *Rice Innovations: Indian Institute of Rice Research*, Hyderabad, 2015, http://www.rkmp.co.in/sites/default/files/Rice%20Innovations2015. pdf Accessed on July 17.

Kato, Y., & Okami, M., (2011). Root morphology, hydraulic conductivity and plant water relations of high-yielding rice grown under aerobic conditions. *Ann. Bot., 108*, 575–583.

Kondo, M., Pablico, P. P., Aragones, D. V., Agbisit, R., Abe, J., Morita, S., & Courtois, B., (2003). Genotypic and environmental variations in root morphology in rice genotypes under upland field conditions. *Plant Soil, 255,* 189–200.

Liang, Y. C., Hu, F., Yang, M. C., & Yu, J. H., (2003). Antioxidative defenses and water deficit-induced oxidative damage in rice (Oryza sativa L.) growing on non-flooded paddy soils with ground mulching. *Plant Soil, 257*(2), 407–416.

Matsuo, N., & Mochizuki, T., (2009). Growth and yield of six rice cultivars under three water-saving cultivations. *Plant Production Science, 12,* 514–525.

Pandian, B. J., (2013). *Tamil Nadu Agricultural University Experiments With Drip Irrigation in Paddy Cultivation.* Press release from the Times of India, Bombay, p. 1.

Pandiyarajan, P., (2016). *Drip Irrigation Helps Paddy Grow in Sodic-Affected Soil.* Press release from the Hindu, New Delhi, p. 2.

Peng, S. B., Bouman, B. A. M., & Visperas, R. M., (2006). Comparison between aerobic and flooded rice in the tropics: Agronomic performance in an eight-season experiment. *Field Crops Res., 96,* 252–259.

Prabhudeva, D. S., Kombali, T. G., Noorasma, S., Giriyappa, M., Thimmegowda, H. D. C. M., & Patil, B., (2016). Precision nutrient management through drip irrigation in aerobic rice. In: *Proceedings of the 13ᵗʰ International Conference on Precision Agriculture by International Society of Precision Agriculture,* Cuba, pp. 23–30.

Qin, J. T., Hu, F., Li, H. X., Wang, Y., & Huang, F., (2006). Effects of non-flooded cultivation with straw mulching on rice agronomic traits and water use efficiency. *Rice Sci., 13*(1), 59–66.

Rekha, B., Jayadeva, H. M., Nagaraju, G. K., Mallikarjuna, G. B., & Geethakumari, A., (2015). Growth and yield of aerobic rice grown under drip fertigation. *The Ecoscan, 9*(1 & 2: Supplement on Rice), 435–437.

Tao, H. B, Holger, B., & Klaus, D., (2006). Growth and yield formation of rice (*Oryza sativa* L.) in the water-saving ground cover rice production system (GCRPS). *Field Crops Res., 95,* 1–12.

The Hindu, New Delhi, (2013). Where Paddy Requires Only Drops of Water. Retrieved on 24ᵗʰ May.

Theivasigamani, P., Vanitha, K., Mohandass, S., Vered, E., Meenakshi, V., Selvakumar, D., Surendran, A., & Lazarovitch, N., (2017). Effect of drip irrigation on growth, physiology, yield and water use of rice. *Journal of Agricultural Science, 9*(1), 1–5.

TNAU, *Brochure on Drip Fertigation in Rice,* (2013). Unpublished Paper at International Research Conference by Tamil Nadu Agricultural University, Coimbatore, 17–19.

Uphoff, N., & Randriamiharisoa, A., (2002). Reducing water use in irrigated rice production with the Madagascar System of Rice Intensification (MSRI). In: Bam Bouman, (ed.), *Water-wise Rice Production* (pp. 71–88). IRRI, Los Baños.

Vijaykumar, P., (2009). *Optimization of Water and Nutrient Requirement for Yield Maximization in Hybrid Rice Under Drip Fertigation System.* M.Sc. (Agri.) Thesis for Tamil Nadu Agricultural University, Coimbatore, p. 140.

Zhang, Z. C., Zhang, S. F., & Yang, J. C., (2008). Yield, grain quality and water use efficiency of rice under non-flooded mulching cultivation. *Field Crops Res., 108,* 71–81.

CHAPTER 10

PERFORMANCE OF FERTIGATED AEROBIC RICE UNDER DRIP IRRIGATION

S. K. NATARAJAN, V. K. DURAISAMY, and K. S. USHARANI

ABSTRACT

The present investigation was carried out to study the influence of irrigation levels and nitrogen doses on aerobic rice under drip irrigation in sandy loam soils. Field experiments were conducted at Agricultural Research Station, Bhavanisagar, Tamil Nadu, India. The treatments included four irrigation levels (irrigation at 100%, 125%, 150% PE daily and conventional irrigation at IW/CPE = 1.25) and three fertigation levels (100,150 and 200 kg N ha^{-1}) of Nitrogen. The experiment was laid in Split-plot design replicated thrice with the test variety as PMK 3, with a duration of 130–135 days. Irrigation was given daily based on daily Pan Evaporation rate. With reference to different irrigation levels, 150% PE on daily basis recorded significantly higher grain yield (5069 kg ha^{-1}), WUE (7.37 kg/ha-mm) and net income of Rs. 33607 ha^{-1} (560.12 US$/ha) and B:C ratio of 1.88. For nitrogen levels, 150 kg N per ha recorded significantly higher grain yield (4146 kg ha^{-1}), WUE (6.69 kg/ha-mm) and net income of Rs. 20464 ha^{-1} (341.07 US$/ha) and B:C ratio of 1.53. For aerobic rice, the irrigation at 150% PE on daily basis combined with 150 kg N per ha recorded significantly higher grain (5483 kg ha^{-1}), WUE (8.18 kg/ha-mm) and higher net income of Rs. 39448 ha^{-1} (657.47 US$/ha) and B:C ratio of 2.03.

10.1 INTRODUCTION

In India, rice is the principal food crop grown in an area of 44.1 million ha with a production of 105.5 million tons and productivity of 2.39 t ha^{-1}. In

Tamil Nadu, it is grown in an area of 1.795million ha with a production of 5.728million tons and productivity of 3191 kg ha⁻¹ (Ministry of Agriculture, 2015). Lowland rice requires around 1000 to 5000 liters of water for producing one kg grain which is about twice or even more than wheat or maize water requirement (Cantrell et al., 2005). However, the increasing scarcity of fresh water for agriculture and the equal demand from the non-agricultural sector threaten the sustainability of the irrigated rice ecosystem. One of the recent developments is to grow rice as an upland crop *viz.* wheat or maize and named as 'aerobic' cultivation. Aerobic rice cultivation saves water input and increases water productivity by reducing water use during land preparation and limiting seepage, percolation, and evaporation (Peng et al., 2015). To make aerobic rice successful, new varieties and management practices must be developed. Optimum irrigation scheduling and nitrogen are critical for profitable yield realization of aerobic rice (Maheswari et al., 2007). Drip irrigation and fertigation methods are water and nutrient efficient methods, respectively in most of the crops apart from increasing the productivity. Information is not available on the response of aerobic rice to drip irrigation and fertigation.

This research study was carried out to study the influence of irrigation levels and nitrogen doses on aerobic rice under drip irrigation in sandy loam soils.

10.2   MATERIALS AND METHODS

Field experiments were conducted during 2013–14 and 2014–15 at Agricultural Research Station, Bhavanisagar, Tamil Nadu. The soil at the experimental site is sandy loam, low in organic carbon (0.46), medium in available phosphorus (21.88 kg ha⁻¹), low in available nitrogen (268 kg ha⁻¹) and high in available potassium (454 kg ha⁻¹). The values of bulk density, particle density, and pore space were 1.27 Mg m⁻³, 1.86 Mg m⁻³ and 31.32%, respectively. The experiment was laid out in a split plot design with four irrigation levels as main plots and three nitrogen doses as subplots. The irrigation and Nitrogen levels were replicated thrice and consisted of:

Main plot: Irrigation levels

M_1: Irrigation at 100% PE daily,
M_2: Irrigation at 125% PE daily,

M_3: Irrigation at 150% PE daily, and
M_4: Conventional irrigation (at IW/CPE = 1.25).

Subplot: Nitrogen levels

S_1: 100 kg N ha^{-1},
S_2: 150 kg N ha^{-1}, and
S_3: 200 kg N ha^{-1}.

Rice variety 'PMK 3' (Paramakudi) of 130–135 days duration was sown by dabbling in raised beds following 20 cm x 10 cm spacing. The Biofertilizer *Azophosmet* @ 2.0 kg ha^{-1} was applied as soil application and seed treatment with *Azophosmet* @ 2 gm kg^{-1} of seeds. Two common irrigations of 60 mm each were given, one at pre-sowing for good germination and second at 10th day after sowing for crop establishment. Thinning and gap filling was done at 14 days after sowing. The blanket fertilizer recommendation of 150:50:50 kg N, P_2O_5 and K_2O per ha were followed besides the basal application of 25 kg Zn SO$_4$ ha^{-1}. The entire quantity of phosphorus and 50% of potassium were applied as basal dose. Different doses of nitrogen and the remaining 50% of potassium were being applied as fertigation in weekly intervals from 21 days after sowing as per the treatment schedule.

Drip irrigation was laid out with 1.5m lateral spacing with 30 cm of dripper spacing with a discharge rate of 8 lph. Irrigation was given daily based on daily Pan Evaporation rate. Pre- and post-harvest observations of growth and yield parameters were recorded following standard procedures. The recorded data were analyzed statistically to find out the significance of the treatment. Net return (Rs. ha^{-1}) was calculated by deducting the cost of cultivation (Rs. ha^{-1}) from the gross returns (Rs. ha^{-1}) excluding the cost incurred towards the installation of a drip system. The other recommended cultural and pest management practices were as recommended by Tamil Nadu Agricultural University.

10.3 RESULTS AND DISCUSSION

The grain yield in any crop is dependent on the photosynthetic source it can build up. A sound source in terms of plant height, number of tillers to support and the number of leaves are logically able to increase the total dry matter and later lead to higher grain yield. Partitioning of dry matter production and its distribution in different parts is important for the determination of the total yield of the crop (Donald, 1962).

In this study, irrigation at 150% PE resulted in significantly taller plants (122.9 cm) compared to irrigation levels at 100% PE, 125% PE and conventional irrigation at IW/CPE =1.25 (Table 10.1). Similarly, a significantly higher number of tillers were recorded with irrigation at 150% PE over the other two irrigation levels at 60 days after sowing (DAS). The number of productive tillers/m² was significantly higher with irrigation at 150% PE compared to irrigation levels at 100% PE, 125% PE and conventional irrigation. However, the 1000 grain weight (g) did not differ significantly among the different irrigation levels. The above results on plant growth and yield attributes were in accordance with Maheswari et al. (2007); Ghosh et al. (2012); Mahajan et al. (2012) and Sridharan and Vijayalakshmi (2012).

TABLE 10.1   Influence of Irrigation Levels and Nitrogen Doses on Growth and Yield Parameters of Aerobic Rice under Drip Irrigation

Treatment	Plant height at harvest (cm)	Tillers/m² at 60 DAS	Productive tillers/m²	1000 grain weight (g)
Irrigation levels (M)				
M_1 – Irri. @ 100 % PE	116.4	393	339	24.2
M_2 – Irri. @ 125 % PE	116.6	385	300	24.5
M_3 – Irri. @ 150 % PE	122.9	519	413	24.6
M_4 – Con.irri.	112.8	409	376	24.4
(IW/CPE =1.25)				
SEd	2.5	39	18	0.1
CD at 5 %	6.0	NS	40	NS
Nitrogen levels (S) (kg/ha)				
S_1 – 100	115.2	397	326	24.1
S_2 – 150	117.7	469	394	24.7
S_3 – 200	118.6	414	351	24.4
SEd	1.2	24	13	0.1
CD at 5 %	2.6	NS	27	NS

Among the nitrogen levels, application of 200 kg N ha⁻¹ resulted in taller plants over 100 kg N ha⁻¹, which in-turn was significant over 150 kg N ha⁻¹. A significant number of tillers/m² and productive tillers/m² were recorded with nitrogen level at 150 kg N ha⁻¹ over the other two nitrogen levels (Figure 10.1). These results are in agreement with findings by Latheef (2010) and Rani (2012). Interaction effects between the irrigation and nitrogen levels with respect to growth and yield attributes by aerobic rice were not significant. Similar findings were reported by Devi and Sumathi (2011) and Reddy et al. (2013).

The grain yield (5069 kg ha⁻¹) of aerobic rice recorded with the irrigation at 150% PE was significantly higher than the other three irrigation levels *i.e.,* 100% PE, 125% PE and conventional irrigation. It was increased by 24% with the 150% PE over 125% PE and 62% over 100% PE respectively (Table 10.2). Comparatively lower grain yield under conventional irrigation condition with soil application of nutrients might be attributed to decrease in synthesis of metabolites and reduction in absorption and translocation of nutrients from the soil to plant. The physiological response of plants by decreased cell division and cell elongation under moderate moisture stress at wider irrigation intervals might have also contributed to reduced grain yield (Sundrapandiyan, 2012). The difference in straw yield due to different irrigation levels was also significant. This is in conformity with Gururaj (2013) and Balaji et al. (2015) in rice.

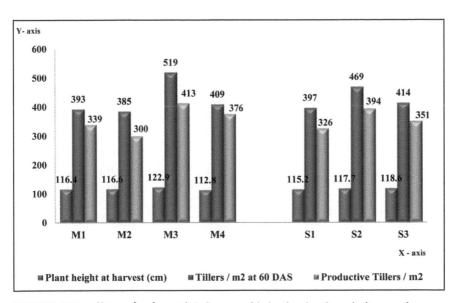

FIGURE 10.1 (See color insert.) Influence of irrigation levels and nitrogen doses on plant height and tillers/m².

Scheduling the irrigation through drip system at 150% PE resulted in 6, 10, and 66% increase in water use efficiency (WUE) over 125% PE, 100% PE and conventional irrigation, respectively due to less water input in the former treatment. The net returns were also higher with the irrigation level of 150% PE (33,607 Rs. ha^{-1} (560.12 US$/ha) compared to 125% PE, which was again better than conventional irrigation. Maheswari et al. (2007) and Reddy et al. (2013) have reported increased yields of aerobic rice with increased frequency and input of water i.e., at 1.2 IW/CPE ratio compared with micro-sprinkler irrigation.

TABLE 10.2   Yield, WUE, Net Return, and B:C Ratio of Aerobic Rice as Influenced by Irrigation Levels and Nitrogen Doses Under Drip Irrigation

Treatment	Grain yield (kg ha^{-1})	Straw yield (kg ha^{-1})	WUE (kg/ ha-mm)	Net return Rs. ha^{-1} (US$/ha)	B:C ratio
Irrigation levels (M)					
M$_1$ – Irri. @ 100 % PE	3137	4389	6.73	6213 (103.55)	1.16
M$_2$ – Irri. @ 125 % PE	4076	5762	6.98	19549 (325.82)	1.51
M$_3$ – Irri. @ 150 % PE	5069	7094	7.37	33607 (560.12)	1.88
M$_4$ – Con.irri. (IW/CPE =1.25)	3057	4910	4.43	11418 (190.30)	1.30
SEd	93	142			
CD at 5 %	228	348			
Nitrogen levels (S) (kg/ha)					
S$_1$ _ 100	3667	5133	5.93	14342 (239.03)	1.38
S$_2$ – 150	4146	5804	6.69	20464 (341.06)	1.53
S$_3$ – 200	4028	5679	6.51	18284 (304.73)	1.47
SEd	62	86			
CD at 5 %	131	183			
Interaction					
SEd	137	200			
CD at 5 %	312	458			

Nitrogen doses applied through drip irrigation, i.e., fertigation differed among themselves with respect to grain and straw yields of aerobic rice (Table 10.2 and Figure 10.2). Application of 150 kg N ha^{-1} significantly increased the grain (4146 kg ha^{-1}) and straw yield (5804 kg ha^{-1}) of aerobic rice over 100 kg N ha^{-1} but at on par with 200 kg N ha^{-1}. Application of fertilizer nutrients through irrigation systems (fertigation) has been reported to increase grain yield (Soman, 2012). Similarly, 150 kg N ha^{-1} resulted in an improvement of 3 and 13% in WUE, over 200 and 100 kg N ha^{-1}, respectively. The net returns were also increased by Rs. 2,180/- per ha (9.34 US$/ha) at 150 kg N ha^{-1} compared to 200 kg N ha^{-1}. A highest benefit-cost ratio was recorded at 150% PE on daily basis (1.88) and 150 kg N ha^{-1} (1.53).

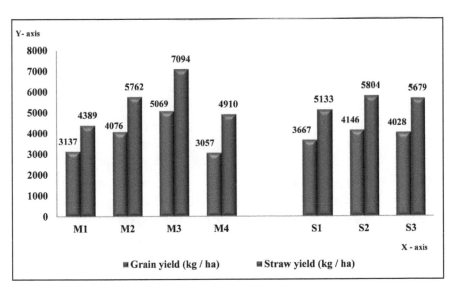

FIGURE 10.2 **(See color insert.)** Grain and straw yield as influenced by levels of irrigation and fertigation.

The combined effect of irrigation levels with nitrogen levels on yield, WUE, net returns, and B:C ratio of aerobic rice under drip irrigation is presented in Table 10.3. The combined effect of irrigation at 150% PE with 150 kg N ha^{-1} (M_3S_2) recorded significantly higher grain yield (5483 kg ha^{-1}), WUE (8.18 kg/ha-mm), net returns (Rs. 39,448 ha^{-1} (657.47 US$/ha) and B:C ratio (2.03), followed by irrigation at 150% PE with 200 kg N ha^{-1} and irrigation at 125% PE with 150 kg N ha^{-1}. The lowest WUE was observed for the crop under conventional irrigation and with the application of different nitrogen levels of 100, 150 and 200 kg N ha^{-1} (4.37, 4.50 and 4.42 kg/ha.mm) (Figure 10.3). The favorable effect of water and nutrients on crop growth and yield in drip irrigation and fertigation probably resulted in higher water use efficiency. Similar results were noticed by Sudhakar et al. (2003) and Lakshmibai et al. (2013).

TABLE 10.3   Yield, WUE, Net-Return, and B:C Ratio of Aerobic Rice as Influenced by the Combined Effect of Irrigation Levels with Nitrogen Doses under Drip Irrigation

Treatment	Grain yield (kg ha^{-1})	Straw yield (kg ha^{-1})	WUE (kg/ha-mm)	Net return Rs. ha^{-1} (US$/ha)	B:C ratio
M_1S_1	3289	4604	7.06	8984 (149.73)	1.24
M_1S_2	3084	4319	6.61	5408 (90.13)	1.14
M_1S_3	3037	4245	6.53	4246 (70.77)	1.11
M_2S_1	3556	4976	6.17	12859 (214.32)	1.34
M_2S_2	4447	6226	7.48	24767 (412.78)	1.65
M_2S_3	4224	6083	7.30	21020 (350.33)	1.54
M_3S_1	4367	6117	6.13	24240 (404.00)	1.64
M_3S_2	5483	7659	8.18	39448 (657.47)	2.03
M_3S_3	5356	7506	7.79	37134 (618.90)	1.96
M_4S_1	3457	4834	4.37	11285 (188.08)	1.30
M_4S_2	3570	5013	4.50	12234 (203.90)	1.31
M_4S_3	3495	4884	4.42	10736 (178.93)	1.28

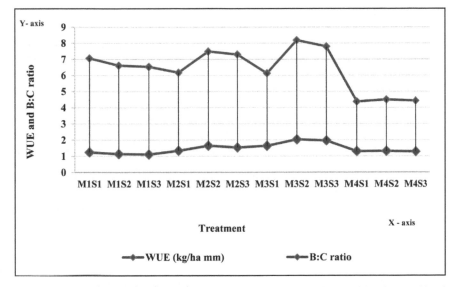

FIGURE 10.3   (See color insert.) WUE and B:C ratio as influenced by the combined effect of irrigation levels with nitrogen doses.

10.4 SUMMARY

The results of the present study showed that higher yield and economic returns were obtained in the western zone of Tamil Nadu in aerobic rice under drip irrigation at 150% daily pan evaporation level and fertigation of 150 kg N ha⁻¹in weekly intervals from 21 days after sowing.

KEYWORDS

- aerobic rice
- drip irrigation
- fertigation
- irrigation levels
- nitrogen levels
- water productivity

REFERENCES

Balaji, N. D., Krishna, M. R., & Pushpa, K., (2015). Yield and yield components of aerobic rice as influenced drip fertigation. *International Journal of Science and Nature, 6*(3), 362–365.

Cantrell, R. P., & Hettel, G. P., (2005). Research strategy for rice in the 21st century. In: Toriyama, et al., (eds.), *Rice is Life: Scientific Perspectives for the 21st Century* (p. 152). Proceedings of the World Rice Research Conference, Tokyo and Tsukuba, Japan.

Devi, M. G., & Sumathi, V., (2011). Effect of nitrogen management on growth, yield and quality of scented rice under aerobic conditions. *Journal of Research ANGRAU, 39*(3), 81–83.

Donald, C. M., (1962). In search of yield. *Journal of the Australian Institute of Agricultural Science, 28*, 171–178.

Ghosh, A., Dey, R., & Singh, O. N., (2012). Improved management alleviating impact of water stress on yield decline of tropical aerobic rice. *Agronomy Journal, 104*(3), 584–588.

Gururaj, K., (2013). *Optimization of Water and Nutrient Requirement Through Drip Fertigation in Aerobic Rice.* M.Sc. Thesis for University of Agricultural Sciences (UAS), Bangalore, India, p. 168.

Lakshmibai, K. J., Ramana, M. K. V., & Venku, N. M., (2013). Effect of graded levels and time of nitrogen application on nutrient uptake, yield and economics of semi-dry rice (*Oryza sativa* L.). *Journal of Research ANGRAU, 41*(2), 21–25.

Latheef, P. M. D., (2010). *Performance of Aerobic Rice Under Different Levels of Irrigation, Nitrogen and Weed Management.* M.Sc. Thesis for Acharya, N. G. Ranga Agricultural University (ANGRAU), Hyderabad, India, p. 205.

Mahajan, G., Chauhan, B. S., Timsina, J., Singh, P. P., & Singh, K., (2012). Crop performance and water and nitrogen use efficiencies in dry-seeded rice in response to irrigation and fertilizer amounts in Northwest India. *Field Crops Research, 134*, 59–70.

Maheswari, J., Maragatham, N., & Martin, G. J., (2007). Relatively simple irrigation scheduling and N application enhances the productivity of aerobic rice *Oryza sativa* L. *American Journal of Plant Physiology, 2*(4), 261–268.

Ministry of Agriculture, Government of India (2014–2015). http://www.indiastat.com. Accessed on July 18, 2018.

Peng, N. L., Bing, S., Chen, M. X., Shah, F., Huang, J. L., Cui, K. H., & Jing, X., (2012). Aerobic rice for water-saving agriculture - A review. *Agronomy for Sustainable Development, 32*(2), 411–418.

Rani, K. S., (2012). *Influence of Nitrogen and Weed Management on Growth and Yield of Aerobic Rice* (*Oryza sativa* L.). M.Sc. Thesis for Acharya N. G. Ranga Agricultural University (ANGRAU), Hyderabad, India, p. 196.

Reddy, M. M., Padmaja, B., Veeranna, G., & Reddy, D., (2013). Response of aerobic rice to irrigation scheduling and nitrogen doses under drip irrigation. *Journal of Research ANGRAU, 41*(2), 144–148.

Soman, P., (2012). Drip fertigation for rice cultivation. In: *Proc. Asian Irrigation Forum* (p. 68). ADB, Manila, Philippines.

Sridharan, N., & Vijayalaxmi, C., (2012). Crop performance, nitrogen and water use in aerobic rice cultivation. *Plant Archives, 12*(1), 79–83.

Sudhakar, G., Solomalai, A., & Ravisankar, N., (2003). Influence of cultivars and levels of nitrogen on yield, nutrient uptake and residual nutrient status of soil in semi-dry rice. *Agricultural Science Digest, 23*(2), 88–91.

Sundrapandiyan, R., (2012). *Study on the Effect of Drip Biogation on the Productivity of Aerobic Rice.* M.Sc. Thesis for Tamil Nadu Agricultural University, Coimbatore, Tamil Nadu, India, pp. 410–423.

PART IV

Drip Fertigated Sugarcane: Management Strategies and Economics

CHAPTER 11

PERFORMANCE OF INTERCROPS UNDER A SUSTAINABLE SUGARCANE INITIATIVE

R. CHANDRASEKARAN, T. SARANRAJ, R. NAGESWARI, C. CHINNUSAMY, and P. DEVASENAPATHY

ABSTRACT

A field experiment was conducted at Sugarcane Research Station, Sirugamani in 2016, to assess the Performance of Intercrops Under Sustainable Sugarcane Initiative. The experiment was laid out in strip plot design with three replications. The main plot treatments comprised of crop geometry. The subplot treatments were intercrop treatment. The intercrops were raised in additive series *viz.*, 3 rows under a row spacing of 150 cm in sugarcane and 4 rows under 180 cm. The recommended schedule of drip fertigation for SSI was followed by using a surface drip irrigation system. The yield parameters of intercrops (like number of pods/plant, pod length, number of seeds/pod, grain yield of black gram and green gram and dry matter production of sunn hemp) were observed under SSI. The results revealed that raising black gram as an intercrop under SSI gave higher yield attributes and grain yield (1286 kg/ha) when it was intercropped with chip budded seedlings of sugarcane in a single row at a row spacing of 150 cm.

11.1 INTRODUCTION

Sugarcane (*Saccharum officinarum*) is emerging as a multi-product crop contributing to the production of sugar, jaggery, alcohol, electricity, paper, and other allied products. The Sustainable Sugarcane Initiative (SSI) is a practical approach to sugarcane production, which is based on the principles of 'more with less' in agriculture like System of Rice Intensification (SRI). SSI improves the productivity of water, land, and labor, while reducing the overall pressure on water resources. SSI is a method of sugarcane production which involves using fewer setts, less water, optimum utilization of fertilizers and land to achieve more yields. In view of dwindling land resources, changing market scenario, consumers' preferences and global competitions, new income generating opportunities need to be created through intercropping in sugarcane.

Sugarcane characteristically widely spaced, initially slow growing, long duration and one-time income generating crop, lends ample scope for intercropping with short duration, high value and mid-season income generating crops for household nutrition and economic security especially of small and marginal cane growers. Intercropping is simultaneous growing of two or more crops with a fixed geometric special arrangement and involves the intensification in both space and time. The productivity of land could be enhanced substantially by growing intercrops between sugarcane rows. Moreover, intercropping in sugarcane is a potent tool to get higher income. Several crops are recommended as intercropping with autumn sugarcane. Under SSI, the farmers have been able to reduce the cost of cultivation by about 21% with increased yield by 24 to 30% apart from extra income from intercrops (NRMC, 2011).

Wider row spacing of 150 cm is preferable for the sugarcane-based intercropping system and both soybean and black gram could be raised as intercrops (Gopalasundaram et al., 2012). Sugarcane planted at 150 cm with intercrops of cowpea, cluster bean, and okra resulted in higher monetary returns with additional income without affecting the main crop. Farmers grow one or more intercrops like cabbage, cauliflower, pulses, marigold etc., along with sugarcane and thereby get an interim income by 90–100 days of planting. When wider row spacing of 150 cm is combined with intercrops like pulses, it increases the cane yield by 20–30 tons/ha and net returns (Rajula-Shanthy et al., 2012). Adoption of integrated nutrient management practices resulted in a reduction of total fertilizer cost with increased productivity and thereby additional income to the practicing farmers (Rajula-Shanthy et al., 2015).

Intercropping is the most efficient and profitable production system especially for small farmers with limited land and inputs resources (Bajwa et al., 1992; Rajula-Shanthy et al., 2012). It is documented that intercropping in sugarcane with short duration crops is agronomically advantageous and could provide additional income (Khippal et al., 2007). Intercropping in sugarcane has been evaluated by many research works. But, the effect of intercrops and their mutual relationship with sugarcane under Sustainable Sugarcane Initiative in Cauvery Deltaic region of Trichy district are only meager.

This chapter focuses on the performance of intercrops under SSI.

11.2 MATERIAL AND METHODS

A field experiment was laid out during *the special* season of 2016 at Sugarcane Research Station, Sirugamani, located at Cauvery delta zone of Tamil Nadu. The geographical location of the experiment site is 10^0 56'N latitude and 78^0 26'E longitude with an altitude of 78.12m above the MSL. The farm receives the total average rainfall of 730.3 mm. The soil at experimental site was well-drained clay loam soil in texture with low available nitrogen, medium in available phosphorus and high in available potassium. The soil analysis indicated 234, 15.8 and 467 kg/ha, respectively of $KMno_4$-N, Olsen P and NH_4OAc-K. Soil had EC of 0.29 dsm^{-1}, pH of 8.58 and organic carbon of 0.58%. The experiments were laid out in strip plot design with four main treatments and four sub-treatments replicated thrice. The net plot size was 27.00 m^2 (9.0 m x 3.0 m). Short duration pulses of green gram (ADT 3), black gram (VBN5) and sunn hemp (*Crotolaria Juncea*, CO1) maturing in 60–75 days were used for the study. The treatments consisted of:

Sugarcane crop geometry, main plots:

M_1 Single-row planting at a row spacing of 150 cm;
M_2 double row planting at a row spacing of 150 cm;
M_3 single row planting at a row spacing of 180 cm; and
M_4 double row planting at a row spacing of 180 cm.

Intercrops, subplots:

S_1 Only Sugarcane;
S_2 Sugarcane + Green gram;
S_3 Sugarcane + Black gram; and
S_4 Sugarcane + Sunn hemp.

The intercrops were raised in additive series *viz.*, 3 rows under a row spacing of 150 cm in sugarcane and 4 rows under 180 cm. The recommended schedule of drip fertigation for SSI was followed under surface drip irrigation system. The number of pods/plant, pod length, number of seeds/pod, grain yield of black gram and green gram and dry matter production and nitrogen, phosphorus, and potash contents of sunn hemp were observed under SSI. Observation was made at harvest stage of intercrops.

11.3   RESULTS AND DISCUSSION

11.3.1   GROWTH AND YIELD ATTRIBUTE OF GREEN GRAM

Performance of intercrops under SSI system is presented in Tables 11.1 to 11.5. Green gram as intercrop under SSI recorded higher mean values of growth and yield attributes such as plant height, number of pods/plant, pod length, number of seeds/pod and grain yield of (61.61, 20.13, 7.01, 7.13 and 970 kg/ha) intercropping with chip budded seedlings of sugarcane in single row at a row spacing of 150 cm (M_1S_2). This was followed by intercropped with chip budded seedlings of single row spacing of 180 cm (M_3S_2). In contrary to this, a study conducted in South Africa indicated that intercrop in alternate cane inter-rows will reduce costs and competition effect on cane yield, by maintaining a useful profit from the food crop.

TABLE 11.1   Effect of Intercrops Under Sustainable Sugarcane Initiative on Growth and Yield Attributes of Green Gram

Treatments	Plant height (cm)	Number of pods/plant	Pod length (cm)	Number of seeds/pod	Grain yield (kg/ha)
M_1S_2	61.61	20.13	7.01	7.13	970
M_2S_2	62.69	20.60	6.67	8.20	854
M_3S_2	63.39	19.87	6.71	8.20	658
M_4S_2	61.27	18.40	6.37	7.87	789
Mean	62.24	19.75	6.69	7.85	970

TABLE 11.2 Effect of Intercrops under Sustainable Sugarcane Initiative on Growth and Yield Attributes of Black Gram

Treatments	Plant height (cm)	Number of pods/plant	Pod length (cm)	Number of seeds/pod	Grain yield (kg/ha)
M_1S_3	61.60	22.80	5.27	4.67	1286
M_2S_3	62.69	28.33	5.02	6.53	1260
M_3S_3	63.38	20.53	5.07	6.73	1214
M_4S_3	61.27	14.73	5.00	5.67	1125
Mean	62.24	21.60	5.09	5.90	1286

TABLE 11.3 Dry Matter Production, N, P & K Accumulation of Sunn Hemp as Intercrop Under SSI

Treatments	DMP (Kg/ha)	Nitrogen (Kg/ha)	Phosphorus (Kg/ha)	Potash (Kg/ha)
M_1S_3	4002	92.00	20.01	72.03
M_2S_3	2008	46.18	10.04	36.14
M_3S_3	2595	59.68	12.97	46.71
M_4S_3	2953	67.91	14.76	53.15
Mean	2889.5	66.44	14.44	52.01

TABLE 11.4 Effect of Intercrops on Germination and Plant Height of Sugarcane Under SSI

Treatments	Germination % at 30 DAP					Plant height at 120 DAP				
	M1	M2	M3	M4	Mean	M1	M2	M3	M4	Mean
S_1	97	85	98	95	94	211.23	218.37	209.33	207.43	211.59
S_2	93	89	93	97	93	175.58	164.95	167.15	175.38	170.77
S_3	94	93	97	86	93	186.46	190.07	187.17	193.36	189.26
S_4	85	91	82	91	87	201.28	205.79	200.40	202.02	202.37
Mean	92	90	93	92		193.64	194.79	191.01	194.55	
	SEd			CD at 0.05		SEd			CD at 0.05	
M	2.9			NS		5.7			NS	
S	3.2			NS		19.4			NS	
M at S	6.6			NS		10.7			NS	
S at M	6.7			NS		21.5			NS	

TABLE 11.5   Effect of Intercrops on Tiller Production ('000/ha) of Sugarcane Under SSI System

Treatments	No. of Tiller at 90 DAP ('000/ha)					No. of Tiller at 120 DAP ('000/ha)				
	M1	M2	M3	M4	Mean	M1	M2	M3	M4	Mean
S_1	60.38	75.84	53.34	56.18	**61.44**	68.64	85.93	63.46	62.72	**70.18**
S_2	66.71	75.61	50.38	60.72	**63.36**	73.46	84.20	59.26	66.67	**70.89**
S_3	55.60	67.88	41.58	51.28	**54.09**	73.09	90.12	55.06	66.05	**71.08**
S_4	79.85	86.88	55.60	66.34	**72.17**	83.83	90.25	63.33	71.23	**77.16**
Mean	65.63	76.55	50.22	58.63	—	74.75	87.62	60.27	66.66	—
	SEd		CD at 0.05			SEd		CD at 0.05		
M	0.3		0.8			5.1		12.6		
S	3.7		9.1			6.4		NS		
M at S	0.7		1.5			6.9		15.9		
S at M	3.7		9.1			7.9		18.5		

11.3.2   GROWTH AND YIELD ATTRIBUTES OF BLACK GRAM

Black gram as intercrop under SSI recorded significantly higher growth and yield attributes such as plant height, number of pods/plant, pod length, number of seed/ pod and grain yield of 61.60, 22.80, 5.27, 4.6 and 1286 kg/ha, respectively, when intercropped with chip budded seedlings of sugarcane in single row at a row spacing of 150 cm (M_1S_2). This was followed by intercropping with chip budded seedlings of double row spacing of 180 cm (M_3S_3).

11.3.3   DRY MATTER PRODUCTION OF SUNN HEMP

Sunn hemp as an intercrop under SSI recorded significantly higher DMP and Nitrogen, phosphorus, and potash of 4002, 92.00, 20.01 and 72.03 kg/ ha, respectively, when intercropped with chip budded seedlings of sugarcane in single row at a row spacing of 150 cm (M_1S_4). This was followed by intercropping with chip budded seedlings of double row spacing of 180 cm (M_3S_4). It is the consequence of *in situ* incorporation of green manure and further decomposition in building the organic matter content of the soil and uptake of applied nutrients by the crop (Kathiresan et al., 1996).

It is further supported with higher biomass production and nitrogen accumulation (Mahendran et al., 1997).

11.3.4   EFFECT OF PLANT GEOMETRY AND INTERCROPS ON GROWTH OF SUGARCANE

Effect of intercrop of black gram, green gram, and sunn hemp showed no significant influence on germination and plant height at 120 DAP of sugarcane under SSI. Plant geometry and intercrops showed significant difference in tiller production of sugarcane at 90 DAP under SSI.

Planting of chip budded seedlings at 150 cm in double rows (M_2) produced significantly higher number of tillers (76,550 per ha) followed by planting at 150 cm in single row (65,930/ha) at 90 DAP. Significantly higher number of tillers (72,170/ha) was produced by sugarcane when intercropped with sunn hemp (S_2). This was followed by intercropping with green gram, which recorded 63,360 tillers/ha. Intercropping black gram with sugarcane under SSI recorded the lowest number of tillers (54,090/ha) (S_3). The interaction effect of plant geometry and intercrops was significant in influencing the tiller production of sugarcane at 90 DAP. The highest number of tillers of 86,880/ha was recorded under a row spacing of 150 cm with double row planting and intercropped with sunn hemp (M_2S_4)

Plant geometry had significant difference in the tiller production of sugarcane at 120 DAP. Planting of chip budded seedlings at 150 cm in double rows (M_2) produced significantly higher number of tillers (90,250/ha) followed by planting at 150 cm in single row (83,830/ha) at 120 DAP. The various intercrops did not influence the tiller production of sugarcane at 120 DAP. The interaction effect of plant geometry and intercrops was significant in influencing the tiller production of sugarcane at 120 DAP. The highest number of tillers of 90,250/ha was recorded under a row spacing of 150 cm with double row planting and intercropped with sunn hemp (M_2S_4). The results agree with those by Nazir et al. (1988).

11.4   SUMMARY

The results revealed that raising black gram as intercrop under SSI gave higher yield attributes and grain yield when intercropped with chip budded seedlings of sugarcane in single row at a row spacing of 150 cm. It is concluded that black gram is a better intercrop under SSI, when sugarcane is planted in single row at row spacing of 150 cm. Planting of chip budded seedlings of sugarcane at a row spacing of 150 cm with double row along with intercropping of sunn hemp recorded significantly highest tiller production in sugarcane under SSI.

KEYWORDS

- **intercropping**
- **sugarcane**
- **sustainable sugarcane initiative**

REFERENCES

Bajwa, A. N., Nazir, M. S., & Mohsin, S., (1992). Agronomic studies on some wheat based intercropping systems. *Pak. J. Agric. Sci., 29*, 439–443.

Gopalasundaram, P., Bhaskaran, A., & Rakkiyappan, P., (2012). INM in sugarcane. *Sugar Tech, 14*(1), 3–20.

Kathiresan, G., & Ayyamperumal, A., (1996). Effect of green manures in intercrops under different sowing methods and nitrogen levels on cane yield. *Co-operative Sugars, 28*, 126–128.

Khippal, A., Marchand, R., Singh, S., & Singh, R., (2007). Intercropping of chickpea in sugarcane with bed planter. In: *Proceedings of 68ᵗʰAnnual Convention of STAI Held at Goa* (pp. 231–237). India, The Sugar Technologists' Association of India (STAI), New Delhi.

Mahendran, S., Porpavai, S., Karamathullah, J., & Ayyamperumal, A., (1997). Effect of green manures on the yield and quality of plant and ratoon cane crop under reduced level and time of application. *Bharatiya Sugar, 22*, 41–44.

Nazir, M. S., Faqeer, I. A., Ali, G., Ahmad, R., & Mahmood, T., (1988). Studies on planting geometry and intercropping in autumn sugarcane. *Gomel. Univ. J. Res., 8*, 57–64.

NRMC, National Resource Management Centre, (2011). *e-Newsletter, 4*, 1–31.

Rajula-Shanthy, T., & Muthusamy, G. R., (2012). Wider row spacing in sugarcane: A socio-economic performance analysis. *Sugar Tech., 14*, 126–130.

Rajula-Shanthy, T., & Subramaniam, R., (2015). Farmers' perspective on integrated nutrient management in sugarcane. *Indian Res. J. Ext. Edu., 15*(1), 100–106.

CHAPTER 12

WATER-SOLUBLE FERTILIZERS FOR A SUSTAINABLE SUGARCANE INITIATIVE UNDER SUBSURFACE DRIP IRRIGATION WITH FERTIGATION

ANBARASU MARIYAPPILLAI, GURUSAMY ARUMUGAM, and INDIRANI RAMESH

ABSTRACT

This chapter focuses on the evaluation of the use of water-soluble fertilizers for sustainable sugarcane initiative (SSI) under subsurface drip irrigation with fertigation. The highest net return (Rs. 3,13,090 Rs./ha (5218 US$/ ha)) was realized under drip fertigation of 100% RDF with ultrasol, MAP, and urea (F7). The next best treatment in increasing the net return was drip fertigation of 100% RDF with urea, MAP, and SOP up to 120 DAP + with Ultrasol from 121 to 210 DAP (F5). But drip fertigation of 100% RDF (P as basal, N&K through drip as urea and MOP) under subsurface drip irrigation system registered the highest B:C ratio (3.70) owing to its lesser cost of cultivation contributed by lower cost of commercial fertilizers.

12.1 INTRODUCTION

Sugarcane is the major commercial crop cultivated on an area of 0.35 million ha with a total production of 46.7 million tons of sugarcane and 16.23 million tons of sugar per annum in Tamil Nadu. The sugarcane productivity has increased over the last two decades. However, the marginal increase in productivity of cane and sugar recovery must be improved by maximizing yield and quality of sugarcane by adopting balanced fertilization (Baki-yathu et al., 2009). In subsurface drip fertigation, nutrient use efficiency may be more than 90% compared to 40–60% in conventional fertilizer application methods. The amount of fertilizer lost through leaching can be less than 10% in fertigation compared to 50% in case of soil application. Adoption of subsurface drip irrigation (SSDI) system may help to increase the water use efficiency and productivity of crops.

In **Sustainable Sugarcane Initiative** (SSI), 25 to 35 days seedlings raised from single bud chips are transplanted in the main field at wider row spacing. Nursery duration saves water requirement in the main field. Since wider spacing is adopted, it is convenient for laying drip irrigation especially subsurface system, and it will facilitate the use of machineries. It was reported that adopting a wider spacing of 180 cm with dual row planting of setts in drip fertigation system can produce higher NMC, individual cane weight, and cane yield. The main concept of SSI is raising seedlings with single bud chips, transplanting seedlings in wider row spacing, balanced fertilization with possible recycling of organic wastes and intercropping. SSI is a method of sugarcane production, which involves using less seeds and raising nursery using single budded chips. Transplanting of 30 – 35 days old seedlings in the main field helps to maintain the required population. Good establishment results in a good start, which provides healthy and strong basis for a better crop yield (Anbarasu et al., 2017; Mahesh, 2009).

This chapter focuses on the evaluation of use of water-soluble fertilizers for sustainable sugarcane initiative (SSI) under subsurface drip irrigation with fertigation.

12.2   MATERIALS AND METHODS

A field experiment was carried out at AICRP-Water Management Research Block, Department of Agronomy, Agricultural College and Research Institute, Tamil Nadu Agricultural University, Madurai during 2013–14. The soil at the experimental field was sandy clay loam, taxonomically classified as Typic Udic Haplustalf with pH- 7.4, organic carbon – 0.48%, EC – 0.42 dS m^{-1}. Soil samples for analyses were initial soil samples and Post-harvest soil samples from the field. The study was designed in RBD with three replications. The treatments were (Table 12.1):

F_1 surface irrigation with soil application of RDF;

F_2 drip fertigation of 100% RDF (P as basal, N&K through drip as urea and MOP);

F_3 drip fertigation of 100% RDF with urea, MAP & SOP;

F_4 drip fertigation of 75% RDF with urea, MAP, and SOP up to 120 DAP + Ultrasol from 121 to 210 DAP;

F_5 drip fertigation of 100% RDF with urea, MAP, and SOP up to 120 DAP + with Ultrasol from 121 to 210 DAP;

F_6 drip fertigation of 75% RDF with Ultrasol, MAP, and urea;

F_7 drip fertigation of 100% RDF with Ultrasol, MAP, and urea;

F_8 drip fertigation of 75% RDF (50% NPK as basal, balance with Ultrasol, MAP, and urea);

F_9 drip fertigation of 100 % RDF (50% NPK as basal, balance with Ultrasol, MAP, and urea).

Note: The test crop variety Co – 86032 and RDF: 275:62.5:112.5 kg NPK ha^{-1}.

The water-soluble fertilizers were Urea. MOP (White Potash), Ultrasol (9:5:33 % NPK), MAP (Mono Ammonium Phosphate), SOP (Sulphate of Potash) for fertigation and Single super phosphate for basal application. Cost of production and gross returns for all treatments were worked out on the basis of prevailing input costs and price of sugarcane at the time of experimentation. Economics were calculated as per standard procedure.

Water was pumped through 7.5 hp submersible motor and was conveyed to field using PVC pipes of 90 mm after filtering through sand and screen filters. From the main line, water was taken to the field through sub mains of 75 and 63 mm diameter PVC pipes. From the sub main, 16 mm size 15 mill low-cost laterals (drip tap) with discharge rate of 1.29 lph were at a spacing of 1.8 m, the laterals were placed in the center of the trenches at 25cm depth from the surface soil. And the end of laterals was connected to collecting sub main PVC pipe (40mm) (Figure 12.1). The operating pressure was maintained at 0.75 kg cm^{-2}. The subsurface drip irrigation system was well maintained by flushing and cleaning the filters.

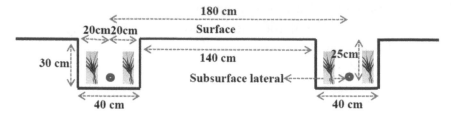

FIGURE 12.1   (See color insert.) Design and layout subsurface drip irrigation with fertigation.

12.2.1 FERTIGATION

The recommended fertilizer dose of 275:62.5:112.5 kg/ha of NPK was followed in the experiment. Fertigation was given as per the treatment schedule. Fertigation was scheduled once in seven days starting from 15 up to 210 DAP. The nutrients were supplied based on the crop growth demand.

The required quantity of N, P, and K fertilizers as urea, Ultrasol, MAP, SOP as per the treatments were dissolved separately in plastic buckets. Required quantity of fertilizer solution was given to each mini fertigation reservoir fixed with each laterals near the sub main and injected through subsurface drip system (Figure 12.2).

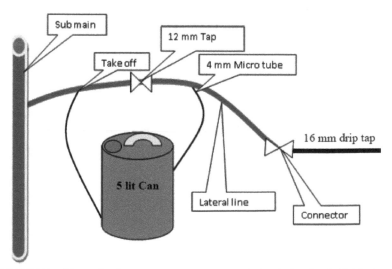

FIGURE 12.2   (See color insert.) Mini fertigation unit for fertigating individual row.

Each plot consisted of five laterals for irrigating five row of sugarcane crop. A tap was provided at the beginning of each lateral for allowing controlled fertigation. Subsurface drip fertigation was carried out in three consecutive steps *viz.,* slightly wetting the root zone before fertigation, fertigating to the field and flushing the nutrients with water.

12.3 RESULTS AND DISCUSSION

12.3.1 CANE YIELD

The results (Table 12.1) clearly indicated positive influence of subsurface drip fertigation levels as well as sources of nutrients on the cane yield. Subsurface drip fertigation of 100% RDF with ultrasol, MAP, and urea (F_7) recorded the maximum cane yield of 175.56 t ha⁻¹, which was followed by drip fertigation of 100% RDF with urea, MAP, and SOP up to 120 DAP + with ultrasol from 121 to 210 DAP (F_5) with the cane yield of 166.21 t ha⁻¹. The lowest cane yield of 107.87 t ha⁻¹ was obtained under surface irrigation with soil application of RDF (F_1).

Subsurface drip fertigation positively influenced the cane yield of SSI. NPK fertigation as WSF through subsurface drip irrigation system boosted the tiller production, recorded higher survival percentage, number of millable canes, cane length, individual cane weight, internode length, grand growth and biological efficiency of the cane (Bill-Segars, 2003; Gaddanaeri et al., 2007). Drip fertigation of 100% RDF with ultrasol, MAP, and urea (F_7) registered significantly higher cane yield (175.56 t ha⁻¹), which amounted to 62.75% yield increase over surface irrigation with soil application of RDF (F_1). It also recorded 33.9% higher cane yield than the fertigation with commercial fertilizers at same level (F_2).

The highest cane yield under subsurface drip fertigation was mainly due to the availability of adequate nutrients and water through the crop growth period. This favorable condition resulted in better and earlier conversion of tillers to millable canes and the early vigor was maintained throughout the crop growth period due to better survival of tillers, which in turn resulted in taller stalks and improved stalk weight at harvest (Khandagave et al., 2005).

The higher cane yield under subsurface drip fertigation compared to conventional method of cultivation in sugarcane was earlier reported by other investigators (Dhotre, 2008; Devi, 2013; Mahesh, 2009).

TABLE 12.1   Effects of Different WSF and Yield of SSI Under Subsurface Drip Irrigation with Fertigation

Treatments		Yield (t ha⁻¹)
F_1	Surface irrigation with soil application of RDF	107.87
F_2	Drip fertigation of 100 % RDF (P as basal, N&K through drip as Urea and MOP)	130.98
F_3	Drip fertigation of 100 % RDF with Urea, Mono-ammonium phosphate (MAP) & sulfate of potash (SOP)	160.39
F_4	Drip fertigation of 75 % RDF with Urea, MAP, and SOP up to 120 DAP + with Ultrasol from 121 to 210 DAP	103.57
F_5	Drip fertigation of 100 % RDF with Urea, MAP, and SOP up to 120 DAP + with Ultrasol from 121 to 210 DAP	166.21
F_6	Drip fertigation of 100 % RDF with Ultrasol, MAP, and Urea	117.23
F_7	Drip fertigation of 75 % RDF (50% NPK as basal, balance with Ultrasol, MAP, and Urea)	175.56
F_8	Drip fertigation of 75 % RDF (50% NPK as basal, balance with Ultrasol, MAP, and Urea)	94.86
F_9	Drip fertigation of 100 % RDF (50% NPK as basal, balance with Ultrasol, MAP, and Urea)	155.95
	SEd	**4.31**
	CD (P = 0.05)	**8.62**

12.3.2   ECONOMICS

The results in Table 12.2 revealed maximum gross income of Rs. 4,65,234 per ha (7754 US$/ha) under drip fertigation of 100% RDF with ultrasol, MAP, and urea (F_7). The minimum gross income of Rs. 2,85,855 per ha (4764 US$/ha) was under surface irrigation with soil application of RDF (F_1).

Recommended level 100% NPK fertigation through WSF under subsurface drip irrigation system recorded higher net income compared to surface irrigation. Among the fertigation treatments, the maximum net income of Rs. 3,13,089 per ha (5218 US$/ha) was realized in fertigation of 100% RDF with ultrasol, MAP, and urea (F_7), whereas the minimum net income of Rs. 1,55,740 per ha (2596 US$/ha) was registered in surface irrigation with soil application of RDF(F_1).

About benefit-cost ratio of SSI cane cultivation under subsurface drip fertigation system, the maximum BC ratio of 3.70 was with drip fertigation of 100% RDF (P as basal; N&K through drip as urea and MOP)-F_2 followed by drip fertigation of 100% RDF with urea, MAP, and SOP. The minimum BC ratio of 2.19 was observed under surface irrigation with soil application of RDF (F_1).

Though drip fertigation of 100% RDF with ultrasol, MAP, and urea(F_7) increased the cost of cultivation, the gross income obtained under this treatment was higher, which was closely followed by drip fertigation of 100% RDF with urea, MAP, and SOP up to 120 DAP with ultrasol from 121 to 210 DAP(F_5).

Drip fertigation of 100% RDF with ultrasol, MAP, and urea (F_7) resulted in higher net return of Rs. 3,13,090 per ha (5218 US$/ha). The next best economically viable treatment was drip fertigation of 100% RDF with urea, MAP, and SOP up to 120 DAP + with ultrasol from 121 to 210 DAP (F_5).

Although the gross and net return were higher under drip fertigation of 100% RDF with ultrasol, MAP, and urea (F_7), yet the B:C ratio (3.06) was numerically lower than (F_2) drip fertigation of 100% RDF (P as basal, N&K through drip as urea and MOP). The high cost of high analytical WSF in addition to the drip system cost resulted in higher cost of cultivation, which ultimately led to lower B:C ratios. The same economic trend has been reported in sugarcane cultivation by other investigators (Devi, 2013; Dhanalakshmi, 1999; Mahesh, 2009; Packiaalakshmi, 2011).

TABLE 12.2 Effects of Different WSFs and Economics of SSI Under SSDF, Rs. per ha or US$/ha

Treatment	Cost of cultivation	Gross income	Net income	BCR
	Rs. per ha or US$/ha			
F_1	130115.27 (2169)	285855.50 (4764)	155740.23 (2596)	2.19
F_2	93770.77 (1563)	347097.00 (5785)	253326.23 (4222)	3.70
F_3	93770.77 (1563)	347097.00 (5785)	253326.23 (4222)	3.70
F_4	88756.68 (1479)	274460.50 (4574)	185703.82 (3095)	3.09

TABLE 12.2   *(Continued)*

Treatment	Cost of cultivation	Gross income	Net income	BCR
		Rs. per ha or US$/ha		
F_5	141319.44 (2355)	440456.50 (7341)	299137.06 (4986)	3.12
F_6	104721.36 (1745)	310659.50 (5177)	205938.14 (3432)	2.97
F_7	152144.42 (2536)	465234.00 (7754)	313089.58 (5219)	3.06
F_8	84842.10 (1414)	251379.00 (4190)	166536.90 (2776)	2.96
F_9	132280.19 (2205)	413267.50 (6888)	280987.31 (4683)	3.12

Note: In this chapter, Rs. 60.00 = US$ 1.00;

Numbers in brackets are in US$ per ha

BCR = [gross income / cost of cultivation]

12.4   SUMMARY

Fertigation through subsurface drip irrigation system is an innovative technology for maximizing the cane yield. Though the unit cost of drip irrigation system was high, considering longer life period of drip irrigation system, the benefit accrued out of drip irrigation will be for longer period. Fertigation with water-soluble fertilizers involved an additional cost. However, the additional cost towards WSF was largely compensated by higher net return obtained by higher yield of sugarcane.

ACKNOWLEDGMENT

The authors are thankful to Coromandel SQM Ltd., for their financial support in this project and special thanks to my advisory committee Dr. Gurusamy Arumugam and Dr. Indirani Ramesh.

KEYWORDS

- **benefit-cost ratio**
- **fertigation**
- **sugarcane**
- **water-soluble fertilizers**

REFERENCES

Anbarasu, M., Gurusamy, A., & Indirani, R., (2017). Enhancing growth and yield attributes of sugarcane sustainable initiative (SSI) through subsurface drip fertigation. *Progressive Research - An International Journal, 12* (Special-I), 1361–1364.

Bakiyathu, B. T., Balaji, T., & Pitchi, J., (2009). *Evaluation of Fertilizer Optima for Improvement of the Quality of Cane in Palaviduthi Soil Series of Tamil Nadu, India.* Tamil Nadu Agricultural University, Coimbatore, p. 250.

Bill-Segars, K., (2003). *Fertigation in IMC Global: Efficient Fertilizer Use Manual, Fourth Edition,* p. 80. http://www.agcentral.com/imcdemo/10fertigation.html Accessed on January 3, 2014.

Devi, P., (2013). *Optimization of Irrigation Regimes and Fertigation Levels for SSI Under Subsurface Drip Fertigation System.* M.Sc. (Ag.) Thesis submitted to Agricultural College and Research Institute, Tamil Nadu Agric. Univ., Madurai, p. 204.

Dhanalakshmi, M., (1999). *Effect of Crop Geometry, Drip and Fertigation on Yield and Quality of Sugarcane.* M.Sc. (Ag.) Thesis submitted to Agricultural College and Research Institute, TNAU, Madurai, Tamil Nadu, p. 188.

Dhotre, R. S., Hadge, S. B., & Rajput, B. K., (2008). Influence of subsurface irrigation through porous pipes on the yield and quality of sugarcane. *J. Maharashtra Agric. Univ., 33*(2), 234–237.

Gaddanakeri, S. A., Kambar-Biradar, P. S., & Nadgouda, B. T., (2007). Response of tillering-sugarcane variety CoC–671 to wider row spacing and clipping. *Karnataka J. Agric. Sci., 20*(3), 598–599.

Khandagave, R. B., Somaiya, S. S., & Hapse, D. G., (2005). Maximization of sugarcane yields and reduction of production costs – participatory rural appraisal. *Proc. Int. Soc. Sugar cane Technol., 25*, 215–218.

Mahesh, R., (2009). *Evaluation of Planting Geometry and Methods of Planting for Sugarcane Under Low Cost Subsurface Drip Fertigation System.* M.Sc. (Ag.) Thesis submitted to Agricultural College and Research Institute, Tamil Nadu Agric. Univ., Madurai, p. 187.

Packiaalakshmi, V., (2011). *Optimization of Nutrient Requirement for Higher Ratoon Sugarcane Productivity Under Subsurface Drip Fertigation System.* M.Sc. (Ag.) Thesis submitted to Agricultural College and Research Institute, Tamil Nadu Agric. Univ., Madurai, p. 198.

CHAPTER 13

PERFORMANCE, WATER PRODUCTIVITY, AND ECONOMICS OF SUGARCANE PRODUCTION IN DIFFERENT AGRO-CLIMATIC ZONES OF TAMIL NADU

B. J. PANDIAN, KANDASAMY VAIYAPURI,
SELVARAJ SELVAKUMAR, and R. CHANDRASEKARAN

ABSTRACT

This research study evaluated the effect of growth, yield, and water productivity of sugarcane cultivation with different methods of cultivation. The results indicated higher plant height at harvest stage under the SSI method (299 cm) compared to conventional planting (262 cm). A number of internodes/plant, number of millable canes/clump, girth, individual cane weight (kg/cane) and cane yield were also superior under SSI method with increased gross income, net return, and B:C ratio both in main crop and ratoon crop. Water consumption was 1820 mm and water productivity was 7.61 kg m^{-3} under the SSI method.

13.1 INTRODUCTION

Sugarcane (*Saccharum officinarum*) is cultivated by 50 million sugarcane farmers in India. Another 5 million people depend on employment generated by 571 sugar factories and related industries. Uttar Pradesh has the largest area (2.302 million ha) under sugarcane followed by Maharashtra (10.52 million ha), Karnataka (0.4 million ha) and Tamil Nadu (0.382 million ha). In India, total production of sugarcane during 2014–15 was 244 M.T.

(Geetha et al., 2015). Over a decade, sugarcane production in India has been fluctuating between 233 and 355 M.T. and the average productivity at the farm level is 40 t ha^{-1}. The production around the world might be reducing by 30% due to climatic changes (Zhao et al., 2015). The crop is facing a rough path ahead due to the increased input and labor cost. Lack of innovative technologies to boost the productivity is another constraint and there are fluctuations in sugarcane productivity. Sustainable Sugarcane Initiative (SSI) can increase the productivity of land, water, and workforce. SSI also aims to reduce the crop duration, in turn, may provide longer crushing period creating employment opportunities for extended duration.

This study was conducted to evaluate the effects of growth, yield, and water productivity of sugarcane cultivation under different methods of cultivation.

13.2   MATERIALS AND METHODS

Field demonstrations were conducted at Western Agro-climatic zone of Tamil Nadu during 2012 to 2014 in main and ratoon crops of sugarcane following SSI and Conventional method with an objective to increase the yield and productivity. Ten demonstrations in Bannari Amman sugar factory zone were conducted over a period of two years and each demonstration was conducted in plots of 0.4 ha. The components of SSI viz., planting of single seedling at wider spacing (5'x2') with drip fertigation were demonstrated in comparison with setts planted under drip irrigation. The soil at the experimental field was alkaline in nature with pH range of 6.5 to 8.34, bulk density 1.23 to 1.27 g cm^{-3} and electrical conductivity 0.28 to 0.31 dSm^{-1}, respectively. The soil texture was clay with 10.75% coarse sand, 33.75% silt and 55% clay with medium depth. The moisture contents at field capacity, permanent wilting point, and available soil moisture were 41.28, 20.27 and 21.01 %, respectively.

The drip irrigation system was installed to meet crop water requirement and for fertigation of water-soluble fertilizers (Table 13.1). Deep plowing with disc plow was followed by operating twice with nine Tyne cultivator across the last plow. Well decomposed FYM @ 12.5 t ha^{-1} was applied at last plowing and mixed with rotovator to obtain fine tilth. Drip laterals were placed at 6" soil depth at spacing of 6 feet. The drip system was operated for 4–8 h based on soil type. Pre-emergence application of atrazine was applied on 3 DAP @ 2.5 kg. All other production technologies were followed as per the TNAU crop production guide (TNAU, 2014).

TABLE 13.1 Fertigation Schedule for Sugarcane (kg ha^{-1})

Days after planting	N	P	K
0–30	39.40	0.00	0.00
31–60	50.60	26.25	9.00
61–90	56.50	20.50	14.50
91–120	60.20	16.25	16.00
121–180	57.80	0.00	40.50
181–210	10.50	0.00	35.0
Total	**275**	**63**	**115**

Thirty days old seedlings var. CO 86032 were planted at wider row spacing of 150 x 60 cm at a depth of 3–5 cm. The main crop was planted during September of 2012. The first ratoon was allowed from the 2nd fortnight of November of 2013 and harvested during September 2014. The quantity of water (liters per day) through drip irrigation was calculated by the climatological approach (Allen et al. 1998) and scheduled on alternate days. In surface irrigation system, the schedule was based on the soil moisture conditions (once in 7–10 days). Plant height at harvest stage, number of internodes per cane, number of millable cane per clump, individual cane weight (kg) and cane yield (kg) were recorded. Economics of cultivation was determined based on the prevailing market price of sugarcane. Water productivity (kg ha^{-3}) was worked out by using the relationship yield /total water consumed.

13.3 RESULTS AND DISCUSSION

13.3.1 GROWTH CHARACTERISTICS

Among the demonstration trials, the trial conducted at Kondapanayakanpudur recorded higher cane plant height (342 cm) under SSI. However, this was at par with the trials conducted at Sokampalayam and Vinnapalli

(Table 13.2). Overall mean value indicated approximately 12% increase in plant height under SSI method. In ratoon sugarcane, SSI method registered higher plant height (293 cm) than conventional planting (248 cm). Growth of sugarcane in terms of plant height was mainly due to wider spacing, more aeration and mother shoot removal on 30–35 DAP. Continuous water and nutrient availability under SSI induced development of more side tillers and enhanced uniform plant growth. Similar results were also observed in ratoon crop. The results corroborated with the findings of Srivastava et al. (1981), who used single bud nursery. Transplanting of Seedling in the main field with wider row spacing utilized the main field with abundant solar radiation, which in turn enhanced tillering, and growth.

13.3.2   YIELD ATTRIBUTES

The internodal length under SSI system registered higher value (13.73 cm) than conventional system (10.95 cm). Number of internodes per plant (26.50), cane girth (9.77 cm), single cane weight (1.81 kg) and number of millable canes per clump (15.12) were higher under SSI method due to continuous supply of nutrient and water, more aeration and easy field operations. In addition, the yield was much higher under SSI method (167 t ha^{-1}) compared to conventional method (138 t ha^{-1}) at Pattanveerthi Ayyanpalayam trial (Table 13.2). Based on the overall mean values, 134 t ha^{-1} was achieved under SSI method of planting compared to 111 t ha^{-1} under CV. (20.17% yield increase).

In ratoon crop, internodal length (cm), number of internodes per plant (no.), cane girth (cm), single cane weight (kg) and millable canes clump^{-1} recorded higher values (13.28 cm, 23.40, 8.76 cm, 1.46 kg and 15.02, respectively) as shown in Table 13.5. Continuous supply of water and inputs at critical stages of crop growth might have increased the vigor and productivity. Among the locations, Annur recorded higher yield under SSI (148 t ha^{-1}), whereas Sathyamangalam location recorded lower yield (121 t ha^{-1}). These results are in conformity with the findings of Singh et al. (2010) and Biksham et al. (2009). Based on the overall mean values of ratoon crop, SSI registered 131.0 t ha^{-1} compared to 104.0 t ha^{-1} under conventional method. This could be possible mainly because of the method of planting, optimum plant population and gap filling under SSI method. The favorable influence on cane weight was due to the supply of required

TABLE 13.2 Growth, Yield Attributes at Harvest and Yield of Sugarcane (Main Crop)

Location	Plant height (cm)		Internode length (cm)		No. of internodes per cane		Girth (cm)		Single cane wt. (kg)		No of millable canes per clump		Cane yield (t ha⁻¹)	
	SSI	CV	SSI	CV	SSI	CV	SSI	CV	SSI	CV	SSI	CV	SSI	CV
Kanoorputhur	316	295	12.80	9.30	26.01	24.02	10.78	8.24	1.50	1.45	11.90	10.47	129	115
Sokampalayam	330	271	14.90	11.20	31.02	22.01	11.45	8.90	1.62	1.39	14.00	12.65	141	119
Vinnapalli	328	289	13.90	10.00	28.03	20.07	9.10	7.70	1.56	1.28	13.30	10.56	127	101
Kondapanayakanpdur	342	285	14.70	10.20	27.04	21.03	9.80	7.20	1.51	1.37	14.70	11.58	138	110
Coimbatore	320	274	12.80	10.10	25.03	19.04	9.90	8.70	1.48	1.36	16.10	10.50	128	103
kembanayakanpalayam	268	239	12.40	11.50	22.10	20.02	9.70	6.70	1.51	1.38	15.40	11.90	124	108
Pattaverthi Ayampalayam	298	274	15.40	12.80	29.20	24.00	10.40	8.90	1.66	1.48	17.50	14.00	167	138
kembanaikanpalayam	216	180	12.56	11.00	20.21	18.00	8.50	6.90	1.32	1.01	15.40	10.50	116	93
Annur	284	235	16.07	13.00	25.12	21.00	8.90	8.20	1.56	1.41	16.10	11.90	131	104
Sathyamangalam	296	278	11.78	10.40	31.03	23.00	9.20	8.50	1.63	1.44	16.80	14.00	158	122
Mean	299	262	13.73	10.95	26.50	21.20	9.77	8.00	1.81	1.47	15.12	11.81	134	111
SEd	6.71		0.45		0.99		0.22		0.04		0.43		2.56	
CD	14.38		0.97		2.13		0.46		0.10		0.92		5.50	

*SSI – Sustainable Sugarcane Initiative; CV – Conventional method.

quantity of water and nutrients at right time and right place as indicated by Loganandhan et al. (2012).

13.3.3   QUALITY PARAMETERS

The overall mean values of Brix (%), polarity (%) and purity (%) were higher under SSI method (18.07, 15.05 and 83.29, respectively) in the main crop (Table 13.3) due to continuous supply of water and nutrient throughout the growth stages of sugarcane and resulting in synchronized maturity of tillers to millable canes. In ratoon crop: the Brix (%), polarity (%) and purity (%) were higher under SSI method (18.31, 15.12 and 82.58, respectively) as shown in Table 13.6. This was mainly due to continuous supply of water and nutrient at peak requirement stages resulting in uniform maturity of tillers to millable canes.

13.3.4   COMMERCIAL CANE SUGAR AND SUGAR YIELD

The overall mean values revealed that SSI registered higher commercial cane sugar recovery and sugar yield (10.38% and 13.99 t ha^{-1}, respectively) compared to CV method (9% and 10.44 t ha^{-1}, respectively) as shown in Table 13.3. The SSI ratoon crop also recorded higher commercial cane sugar recovery and sugar yield (10.38% and 13.59 t ha^{-1}, respectively) as shown in Table 13.6 due to synchronized maturity of tillers to millable canes and appreciable cane quality parameters.

13.3.5   WATER PRODUCTIVITY AND ECONOMICS

Total water consumption and water productivity were determined for both methods of planting. SSI method showed less water consumption (1787 mm) and more water productivity (7.61 kg/m^3) compared to conventional system (1927 mm and 5.82 kg m^{-3}). Gross return, net return, and B:C ratio analysis indicated an additional net return of Rs. 40,610 per ha (676.83 US$/ha) under SSI resulting in a B:C ratio of 1.91 as shown in Table 13.4. In ratoon crop total water consumed was 1787mm as shown in Table 13.7 (but with water productivity was higher under SSI (7.31 kg-m^{-3}). SSI method also recorded more gross return, net return and B:C ratio (Rs.3,00,610

TABLE 13.3 Quality Characters and Sugar Yield of Main Crop

Location	Brix content (%)		Polarity (%)		Purity (%)		CCS (%)		Sugar yield (t ha^{-1})	
	SSI	CV	SSI	CV	SSI	CV	SSI	CV	SSI	CV
Kanoorputhur	18.30	17.60	15.10	13.66	89.62	86.71	10.73	9.91	13.84	11.40
Sokampalayam	18.58	17.48	15.36	13.77	82.67	78.78	10.55	9.21	14.87	10.96
Vinnapalli	17.24	17.43	15.14	13.77	87.82	79.00	10.73	9.23	13.63	9.32
Kondapanayakanpdur	17.80	16.43	14.88	13.07	83.60	79.55	10.28	8.79	14.19	9.70
Coimbatore	18.40	17.43	14.98	13.77	81.41	79.00	10.21	9.23	13.07	9.51
Kembanayakanpalayam	17.92	17.28	14.85	13.62	82.87	78.82	10.22	9.12	12.67	9.85
Pattaverthi Ayampalayam	18.53	17.57	15.21	13.68	82.08	77.86	10.41	9.09	17.38	14.81
Kembanaikanpalayam	17.76	17.36	13.84	12.86	77.93	74.08	9.20	8.30	10.67	7.72
Annur	17.84	17.32	14.65	13.58	82.12	78.41	10.03	9.06	13.14	9.42
Sathyamangalam	18.33	17.00	15.18	13.58	82.82	79.88	10.44	9.16	16.50	11.75
Mean	18.07	16.99	15.05	13.44	83.29	79.21	10.38	9.01	13.99	10.44
SEd	0.33		0.25		0.56		0.17		0.32	
CD	0.70		0.54		1.20		0.38		0.69	

*SSI – Sustainable Sugarcane Initiative; CV – Conventional method; CCS = Commercial cane sugar.

TABLE 13.4 Water Productivity and Economics of Sugarcane Production (Main Crop)

Location	Total water consumed including ER (mm)		Water productivity (kg m⁻³)		Gross return, Rs/ha (US$/ha)		Cost of Cultivation, Rs/ha (US$/ha)		Net return, Rs/ha (US$/ha)		B:C ratio	
	SSI	CV	SSI	CV	SSI	CV	SSI	CV	SSI	CV	SSI	CV
Kanoorputhur	1750	1900	7.37	6.52	296700 (4945)	273700 (4562)	159500 (2658)	157000 (2617)	137200 (2287)	116700 (1945)	1.86	1.74
Sokampalayam	1800	1950	7.83	6.10	324300 (5405)	282900 (4715)	165500 (2758)	159000 (2650)	158800 (2647)	123900 (2065)	1.95	1.78
Vinnapalli	1820	1980	6.98	5.10	292100 (4668)	241500 (4025)	158500 (2642)	150000 (2500)	133600 (2227)	91500 (1525)	1.84	1.61
Kondapanayakanpdur	1780	1870	7.75	5.88	317400 (5290)	261200 (4353)	164000 (2733)	154500 (2575)	153400 (2557)	106700 (1778)	1.93	1.69
Coimbatore	1860	1950	6.88	5.28	294400 (4907)	246100 (4102)	159500 (2658)	151000 (2517)	134900 (2248)	95100 (1585)	1.85	1.63
kembanayakanpalayam	1680	1820	7.38	5.93	285200 (4753)	257600 (4293)	157000 (2617)	161500 (2692)	128200 (2137)	96100 (1602)	1.82	1.60
Pattaverthi Ayampalayam	1780	1960	9.43	7.04	386400 (6440)	326600 (5443)	179000 (2093)	168500 (2808)	207400 (3457)	158100 (2635)	2.16	1.94
kembanaikanpalayam	1740	1900	6.67	4.89	266800 (4447)	223100 (3718)	153000 (2550)	146000 (2433)	113800 (1897)	77100 (1285)	1.74	1.53
Annur	1840	1990	7.12	5.22	301300 (5022)	248400 (4140)	160500 (2675)	151500 (2525)	140800 (2347)	96900 (1615)	1.88	1.64
Sathyamangalam	1820	1950	8.68	6.26	363400 (6057)	289800 (4830)	174000 (2900)	160500 (2675)	189400 (3157)	129300 (2155)	2.08	1.81
Mean	**1787**	**1927**	**7.61**	**5.82**	**312800 (5213)**	**265090 (4418)**	**163050 (2718)**	**155950 (2599)**	**149750 (2496)**	**109140 (1810)**	**1.91**	**1.70**

*SSI – Sustainable Sugarcane Initiative; CV – Conventional method.
Note: In this chapter, 60.00 Rs. = 1.00 US$; Note: Numbers in brackets are in US$.

TABLE 13.5 Growth, Yield Attributes, and Yield of Sugarcane (Ratoon Crop)

Location	Plant height (cm)		Internode length (cm)		No. of inter node per plant (No)		Girth (cm)		Single cane wt. (kg)		No of Millable cane clump⁻¹		Cane yield (t ha⁻¹)	
	SSI	CV	SSI	CV	SSI	CV	SSI	CV	SSI	CV	SSI	CV	SSI	CV
Kanoorputhur	302	276	11.90	10.40	21.25	19.14	7.97	6.24	1.43	1.23	13.30	10.10	120	92
Sokampalayam	336	289	14.00	10.00	21.92	21.43	8.22	5.59	1.48	1.25	13.50	10.40	125	93
Vinnapalli	321	275	12.80	10.20	22.14	19.90	8.47	5.27	1.42	1.16	12.70	10.80	124	98
Kondapanayakanpdur	335	278	14.00	11.00	23.14	19.90	8.55	6.08	1.51	1.19	15.40	12.60	136	112
Coimbatore	318	268	13.50	11.10	21.36	18.37	8.22	5.99	1.45	1.25	14.00	11.20	126	97
kembanayakanpalayam	331	287	13.20	11.00	23.70	22.96	9.96	7.61	1.48	1.26	15.80	11.20	130	105
Pattaverthi Ayampalayam	246	199	15.70	10.20	25.47	17.60	8.20	7.80	1.49	1.38	16.90	15.40	135	121
kembanaikanpalayam	227	189	11.90	10.00	21.36	18.37	8.96	6.32	1.30	1.23	14.20	11.20	118	91
Annur	291	235	13.30	10.20	27.36	20.37	9.96	9.23	1.52	1.27	16.20	12.60	148	116
Sathyamangalam	227	188	12.50	11.50	26.36	18.37	9.13	7.70	1.49	1.32	18.20	14.00	145	120
Mean	293	248	13.28	10.56	23.40	19.64	8.76	6.78	1.457	1.25	15.02	11.95	131	104
SEd	2.88		0.41		0.30		0.28		0.05		0.43		2.56	
CD	6.18		0.88		0.65		0.61		0.12		0.93		5.49	

*SSI – Sustainable Sugarcane Initiative; CV – Conventional method.

TABLE 13.6　Quality Characters and Sugar Yield (Ratoon Crop)

Location	Brix (%)		Polarity (%)		Purity (%)		CCS (%)		Sugar yield (t ha^{-1})	
	SSI	CV	SSI	CV	SSI	CV	SSI	CV	SSI	CV
Kanoorputhur	17.91	17.00	14.80	13.60	82.64	80.00	10.17	9.18	12.20	8.45
Sokampalayam	18.10	17.00	15.40	13.22	85.08	77.76	10.74	8.78	13.42	8.17
Vinnapalli	18.10	17.00	15.20	13.60	83.98	80.00	10.53	9.18	13.06	9.00
Kondapanayakanpdur	18.11	18.11	14.50	14.11	80.07	77.91	9.79	9.38	13.32	10.51
Coimbatore	18.30	17.10	14.90	13.20	81.42	77.19	10.16	8.73	12.80	8.47
kembanayakanpalayam	18.30	18.10	15.40	13.90	84.15	76.80	10.68	9.17	13.88	9.63
Pattaverthi Ayampalayam	18.60	17.70	15.70	14.65	84.41	82.77	10.91	10.07	14.72	12.18
kembanaikanpalayam	18.64	17.90	14.26	13.90	76.50	77.65	9.38	9.23	11.07	8.40
Annur	18.40	17.84	15.64	14.50	85.00	81.28	10.90	9.87	16.13	11.45
Sathyamangalam	18.66	17.89	15.40	14.25	82.53	79.65	10.57	9.60	15.33	11.52
Mean	18.31	17.56	15.12	13.89	82.58	79.10	10.38	9.32	13.59	9.78
SEd	0.12		0.18		0.80		0.17		1.56	
CD	0.27		0.38		1.73		0.36		3.34	

*SSI – Sustainable Sugarcane Initiative; CV – Conventional method; CCS = Commercial cane sugar.

TABLE 13.7 Water Productivity and Economics of Sugarcane under Two Methods of Planting (Ratoon Crop)

Location	Total water consumed including ER (mm)		Water Productivity (kg m⁻³)		Gross return, Rs/ha (US$/ha)		Cost of Cultivation, Rs/ha (US$/ha)		Net return, Rs/ha (US$/ha)		B:C ratio	
	SSI	CV	SSI	CV	SSI	CV	SSI	CV	SSI	CV	SSI	CV
Kanoorputhur	1750	1750	6.86	5.26	276000 (4600)	211600 (3527)	130000 (2167)	116000 (1933)	146000 (2433)	95600 (1593)	2.12	1.82
Sokampalayam	1800	1800	6.94	5.17	287500 (4792)	213900 (3565)	132500 (2208)	116500 (1942)	155000 (2583)	97400 (1623)	2.17	1.84
Vinnapalli	1820	1820	6.81	5.38	285200 (4753)	225400 (3757)	132000 (2200)	119000 (1983)	153200 (2553)	106400 (1773)	2.16	1.89
Kondapanayakanpdur	1780	1780	7.64	6.29	312800 (5213)	257600 (4293)	138000 (2300)	126000 (2100)	174800 (2913)	131600 (2193)	2.27	2.04
Coimbatore	1860	1860	6.77	5.22	289800 (4830)	223100 (3718)	133000 (2217)	118500 (1975)	156800 (2613)	104600 (1743)	2.18	1.88
kembanayakanpalayam	1680	1680	7.74	6.25	299000 (4983)	241500 (4025)	135000 (2250)	122500 (2042)	164000 (2733)	119000 (1983)	2.21	1.97
Pattaverthi Ayampalayam	1780	1780	7.58	6.80	310500 (5175)	278300 (4638)	137500 (2292)	130500 (2175)	173000 (2883)	147800 (2463)	2.26	2.13
kembanaikanpalayam	1740	1740	6.78	5.23	271400 (4523)	209300 (3488)	129000 (2150)	115500 (1925)	142400 (2373)	93800 (1563)	2.10	1.81
Annur	1840	1840	8.04	6.30	340400 (5673)	266800 (4447)	144000 (2400)	128000 (2133)	200400 (3340)	138800 (2313)	2.36	2.08
Sathyamangalam	1820	1820	7.96	6.59	333500 (5558)	276000 (4500)	142500 (2375)	130000 (2167)	191000 (3183)	146000 (2433)	2.34	2.12
Mean	1787	1787	7.31	5.85	300610 (5002)	240350 (4006)	135350 (2256)	122250 (2038)	165660 (2761)	118100 (1968)	2.22	1.96

*SSI – Sustainable Sugarcane Initiative; CV – Conventional method.

Note: Numbers in brackets are in US$.

(5010.17 US$/ha); Rs.1,65,660 (2761.00 US$/ha) and 2.22, respectively) when compared to conventional method (Rs. 2,40,350 (4005.83 US$/ha); Rs.1,18,100 (1968.33 US$/ha) and 1.96, respectively). An additional Rs.13,100 (218.33 US$) was incurred under SSI method, which in turn increased net returns up to Rs.47, 560 (792.67 US$). The results were in conformity with the statement of Vinodgoud (2011), who also showed that SSI is one of the best methods to improve cane productivity with reduced cost of cultivation.

The overall system productivity indicated that SSI could record 265 t ha^{-1} (main + ratoon) compared to 215 t/ha under conventional system as shown in Table 13.8. The economics of production system indicated that SSI had registered higher cost of cultivation as Rs. 2,98,400 (7460.00 US$) per ha. However, due to increased yield and quality canes, more net return Rs 3,15,410 per ha (5256.83 US$/ha) was possible, while, conventional system could give only Rs 2,27,240 per ha (3787.33 US$/ha).

13.4   SUMMARY

This study indicated higher plant height at harvest stage under SSI method (299 cm) compared to conventional planting (262 cm). Other yield attributing characters (*viz.*, number of internodes/plant, number of millable canes/clump, girth, individual cane weight (kg/cane) and cane yield) were also superior under SSI method with increased gross income, net return, and B:C ratio both in main crop and ratoon crop. Water consumption was lower (1820 mm) and water productivity was higher (7.61 kg m^{-3}) in SSI method compared to conventional planting (1927 mm of water consumption and 5.82 kg m^{-3} of water productivity, respectively).

KEYWORDS

- sugarcane
- water consumption
- water productivity
- yield

TABLE 13.8 Economics of Sugarcane Production Under SSI

Method	Yield (Rs ha⁻¹)			Gross return, Rs/ha (US$/ha)			Cost of Cultivation, Rs/ha (US$/ha)			Net return, Rs/ ha (US$/ha)		
	Main crop	Ratoon	Total	Main crop	Ratoon	Total	Main crop	Ratoon	Total	Main crop	Ratoon	Total
SSI	134	131	265	163050	135350	298400	312800	300610	613410	165660	149750	315410
				(2718)	(2256)	(4973)	(5213)	(5010)	(10224)	(2761)	(2496)	0
CV	111	104	215	155950	122250	278200	265090	240350	505440	118100	109140	227240
				(2599)	(2038)	(4637)	(3318)	(4006)	(8424)	(1968)	(1819)	0

*SSI – Sustainable Sugarcane Initiative; CV – Conventional method.
Note: Numbers in brackets are in US$.

REFERENCES

Allen, R. G., & Pereira, L. S., (1998). *Crop Evapotranspiration - Guidelines for Computing Crop Water Requirements*. FAO Irrigation and Drainage Paper 56, FAO, Rome, Italy, p. 213.

Biksham, G., Loganadhan, N., Vinodgoud, V., & Agarwal, M., (2009). *Sustainable Sugarcane Initiative (SSI) – Improving Sugarcane Cultivation in India*. An initiative of ICRISAT-WWF project, ICRISAT, Patancheru - 502324, Andhra Pradesh, India, Training Manual 3, p. 7.

Geetha, P., Sivaraman, K., Tayade, A. S., & Dhanapal, R., (2015). Sugarcane based intercropping system and its effect on cane yield. *Journal of Sugarcane Research, 5,* 1–10.

Loganandhan, N., Biksham, G., Vinodgoud, V., & Natarajan, U. S., (2012). Sustainable sugarcane initiative (SSI): Methodology of 'more with less'. *Sugar Tech., 15,* 98–102.

Singh, G. K., Yadav, R. L., & Shukla, S. K., (2010). Effect of planting geometry, nitrogen, and potassium application on yield and quality of ratoon sugarcane in sub–tropical climatic conditions. *Ind. J. of Agric Sci., 80*(12), 1038–1042.

Srivastava, K. K., Narasimhan, R., & Shukla, S. K., (1981). A new technique for sugarcane planting. *Indian Farming, 31,* 15–17.

TNAU, (2014). *Crop Production Guide*. Tamil Nadu Agricultural University (TNAU), Coimbatore–641003, online.

Vinodgoud, V., (2011). Sustainable sugarcane initiative, SSI - A methodology for improving yields. In: *Proceedings of First National Seminar on Sustainable Sugarcane Initiative, SSI: A Methodology to Improve Cane Productivity* (p. 8). Tamil Nadu Agricultural University, Coimbatore.

Zhao, D., & Rui, Y., (2015). Climate change and sugarcane production: Potential impact and mitigation strategies. *International Journal of Agronomy, Article ID 547386,* p. 10, https://www.hindawi.com/journals/ija/2015/547386 Accessed on August 31, 2018.

CHAPTER 14

ECONOMIC RETURNS OF SUSTAINABLE SUGARCANE INITIATIVE TECHNOLOGY IN SUGARCANE CULTIVATION

V. SARAVANAKUMAR, K. ARTHI, R. BALASUBRAMANIAN, and K. DIVYA

ABSTRACT

The present study was conducted because of significant production and consumption of sugarcane in Tami Nadu. This chapter explores the impact of SSI on resource conservation, energy saving and economic returns in sugarcane cultivation. The results showed that the productivity and profitability of sugarcane cultivation were more under SSI technology than the conventional method. The major sources of productivity enhancement under SSI were fertilizers, drip irrigation, micro-nutrients, and deployment of labor. Under the SSI method, a substantial amount of water (40%) and electricity (55%) consumption was reduced compared to the conventional method. Low procurement price, limited drip irrigation subsidy, clogging in the drip system and less availability of quality seed materials were the major constraints faced by the farmers. The recommended policy options for improving sugarcane production and its profitability are: upscale the adoption of drip irrigation and its subsidy, increase the availability of technical services to remove the clogs and impart periodical training to farmers on SSI technology.

14.1 INTRODUCTION

Sugarcane is an important cash crop, which not only produces 78% of the sugar worldwide, but also contributes to energy demands by cogeneration and alcohol as fuel and used for producing number of value-added products (Shrivastava et al., 2011). About 35 million sugarcane farmers and large number of agricultural laborers are involved in sugarcane cultivation and ancillary activities. Apart from this, the sugar industry provides employment to 50 million people including the employment generated by around 570 sugar factories and other related industries (ISMA, 2015). Today, India ranks second in the world, after Brazil, in terms of area (5.06 m-ha) and sugarcane production (348 million tons in the year 2013–14).

However, sugarcane cultivation in India is in crisis. During the last 10 years, sugarcane production has been fluctuating widely, between 233 and 355 million tons. At the same time, productivity at the farm level has been stagnant over the last two decades, at around 65–70 tons/ha. Sugarcane area declined to 4.6 million ha down by 7% in 2014–15. In the year 2015–2016, India's sugar production declined to 25 million tons, down by 11% from the preceding year.

The cost of production of sugarcane has been increasing due to increase in input costs, especially the cost of labor driven by scarcity of labor due to implementation of Mahatma Gandhi National Rural Employment Guarantee Scheme (MGNREGS). This scarcity of labor has increased the wages. Traditional method of sugarcane production also entails significant environmental costs due to large quantity of water used in sugarcane production. Farmers are less aware of the latest production techniques and farm management strategies, which help to increase output as well as reduce economic and environmental costs.

In India, the sugar industry is also facing problems such as inadequate cane supply for crushing due to reduction in area under sugarcane, labor scarcity for harvesting, competition from other remunerative crops like rice and maize, and inadequate availability of planting material at the time onset of season leading to inadequate coverage of targeted area. The use of untreated and poor-quality planting material used by farmers, has hampered the cane quality which resulted in poor sugar recovery. Further, less mechanization of sugarcane production due to closer spacing in conventional method of planting increased the drudgery of human labor and its cost. Under these circumstances, both farmers and sugar industry are in distress. In addition, cane farmers have switched to other crops due to non-payment or delayed payment of money by the sugar mills. Besides, the expectation of rise in demand has led to some traders hoarding stocks. Therefore. crisis in sugarcane production calls for alternative methods and technologies in sugarcane cultivation to make it viable, and remunerative both farmers and sugar industry and environmentally less damaging.

Apart from these, sugarcane is one of the water-intensive crops, it has been cultivated mainly under surface method of irrigation, where water use efficiency is very low (35–40%) owing to substantial evaporation and distribution losses (Rosegrant, 1997; Rosegrant et al., 1996; Sivanappan, 1994). Water is increasingly becoming a major limiting factor for irrigated crops, especially for sugarcane. Further, the erratic trends in rainfall add to the growing complexity of the water scarcity issues. Thus, we need to explore every possible approach to reduce the water input to all crops, particularly those which excessively depend on scarce resources. Any water reduction to water-guzzling crop such as sugarcane will have a positive impact on the other crops in the same region.

The solution to improve productivity and conserve water resources calls for an integrated approach to agriculture involving all stakeholders. The Sustainable Sugarcane Initiative (SSI) is a step in that direction to address the critical issues facing sugarcane cultivation. SSI provides a viable and sustainable alternative to farmers for improving the productivity of their land, water, and labor, all at the same time. By reducing the overall pressure on water resources SSI contributes to the preservation and recovery of ecosystems. SSI is a set of practices based on the principle of producing 'More with Less' in agriculture. This is a farm-based method and farmers have the option to use the cane variety of their choice. Considering the water scarcity, one of the methods introduced to increase the water use efficiency recently in Indian agriculture is drip method of irrigation (DMI).

This chapter explores impact of SSI on resource conservation, energy saving and economic returns in sugarcane cultivation in Tamil Nadu.

14.2 METHODOLOGY

The present study was conducted in Tamil Nadu because of significant production and consumption of sugarcane in the state. The multistage random sampling technique was followed to select 120 sample households from the districts of Villupuram and Trichy. The primary data were collected during the year 2014–15 through a well-structured interview schedule. To estimate cost and returns of sugarcane under conventional and SSI methods, the standard method developed by the Commission on

Agricultural Costs and Prices (CACP) was followed (Raju et al., 1990). The variable costs cover actual expenses incurred in sugarcane production. The actual cash and kind expenses are: (a) planting materials (setts/ seedlings) (b) manures (c) fertilizers (c) plant protection chemicals and (d) hired human and machine labors. The fixed costs cover depreciation, interest, and rental value of land. Depreciation charges on farm implements, farm machineries and fixed capital such as drip irrigation structures, farm buildings etc., were calculated using the straight-line method.

To measure the sources of change in sugarcane productivity, Cobb-Douglas production function was used as follows:

$$YLD_i = \beta_0 \, NITRO \,^{\beta i1}_{\ i1} \, PHOS^{\beta i2}_{\ i2} \, POTAS^{\beta i3}_{\ i3}$$
$$MNM \,^{\beta i4}_{\ i4} \, HLAB \,^{\beta i5}_{\ i5} \, MLAB \,^{\beta i}_{\ i6} u_i \qquad (1)$$

where: subscript $i = 1$ indicates conventional method; $i = 2$ indicates SSI method; YLD = sugarcane yield (t/ha); $NITRO_{i1}$ = quantity of nitrogenous fertilizers used (N) (kg/ha); $PHOS_{i2}$ = quantity of phosphatic fertilizers used (kg/ha); $POTAS_{i3}$ = quantity of potassic fertilizers used (kg/ha); MNM_{i4} = micronutrients applied (kg/ha); $HLAB_{i5}$ = human labor (person days/ha); $MLAB_{i6}$ = machine labor (hours/ha); β_0 = [intercept-term (scale parameter); and u_i = [error-term independently distributed with zero mean and constant variance]; $\beta_{i1}, \beta_{i2}, \beta_{i3}, \beta_{i4}, \beta_{i5}$ and β_{i6} are the regression coefficients of nitrogenous, phosphatic, and potassic fertilizers, micronutrients, human labor and machine labor, respectively.

The family labor was imputed and evaluated at the prevailing wage rates of hired labor at the village level. Chow's test (Gujarati et al., 2014) was employed to identify whether the parameters governing the production relations were different in the "conventional method" and "SSI method" and it was used to compute the 'F' ratio. The computed 'F' value was compared with 'F' critical value 'p' and $(n + m - 2p)$ degrees of freedom at appropriate level of significance; where, 'n' refers to the number of observations and 'm' refers to the number of variables. The non-significant 'F' value indicated no structural difference between conventional and SSI methods. In this study, the output decomposition model (Bisaliah, 1977) was used to examine the productivity difference between conventional and SSI methods due to technological change and input use.

Garrett's ranking technique was used to prioritize the constraints faced by the farmers in sugarcane cultivation and sugar factories in popularizing the SSI method of sugarcane cultivation (Mohanasundaram, 2015; Shanthy et al., 2014). In the Garrett's ranking technique, the respondents were asked to rank the factors or problems and these ranks were converted into% position. Garrett's ranking was computed using the following formula:

$$\text{Percent position} = [100 \times (R_{ij} - 0.5)]/N_j \qquad (2)$$

where, R_{ij} = Ranking given to the ith attribute by the jth individual; N_j = Number of attributes ranked by the jth individual.

The percent position of each rank was converted into scores by referring to Tables given by Garrett (1969). Then, for each factor, the scores of the various respondents were added and the mean value was estimated.

14.3   RESULTS AND DISCUSSION

14.3.1   ECONOMIC ANALYSIS OF SUGARCANE CULTIVATION UNDER SSI AND CONVENTIONAL METHODS

The economics of sugarcane cultivation under conventional and SSI methods are presented in Table 14.1. The overall total cost of sugarcane cultivation was Rs. 179,000/ha (2983.33 US$/ha) under conventional method and to about Rs. 207,170/ha (3452.83 US$/ha) under SSI method. In this cost, the share of total fixed cost ranged from 17 to 21% and the remaining 79 to 83% was accounted for by the total variable cost. The total fixed cost component (depreciation and interest) incurred in SSI method was more due to additional expenditure on establishment of drip irrigation infrastructure and its maintenance.

Among the variable cost components, the cost of human labor was the highest (50%). The major share of labor cost accounted for harvesting cost, which ranged from 34 to 37%, in both conventional and SSI methods, respectively. The share of wages for harvesting was higher in the SSI method due to harvesting higher cane yield compared to for the conventional method. Expenditure on fertilizers was the second major variable cost, which accounted for approximately 10% of total costs in both the methods. The cost of plant protection chemicals (PPC) was higher in the conventional (Rs. 2915 (48.58 US)) than under SSI (Rs. 2400 (40.00 US)) method. The expenditure incurred on planting materials (setts) was less (Rs. 19,736 (328.93 US)) in conventional than SSI (Rs. 20,994 (349.90 US)) method.

TABLE 14.1   Cost and Returns (Rs. per ha or (US$/ha)) of Sugarcane Cultivation Under Conventional and SSI Methods. Note: Numbers in brackets are percentage of total under each method.

Particulars	Villupuram district		Trichy district		Overall	
	Conventional	SSI	Conventional	SSI	Conventional	SSI
Total Fixed cost (TFC), Rs.	30373	43911	30682	42203	30528	43057
	(16.90)	(21.04)	(17.21)	(20.52)	(17.05)	(20.78)
Variable costs, Rs.						
Setts cost, Rs.	19385	18935	20086	23053	19736	20994
	(10.79)	(9.07)	(11.27)	(11.21)	(11.02)	(10.13)
Fertilizer cost, Rs.	18793	21568	17894	20044	18344	20806
	(10.46)	(10.34)	(10.04)	(9.75)	(10.25)	(10.04)
Plant protection chemicals cost, Rs.	2793	2540	3037	2259	2915	2400
	(1.55)	(1.22)	(1.70)	(1.10)	(1.63)	(1.16)
Human labor Cost. Rs.	93225	106726	91228	102954	92227	104840
	(51.87)	(51.14)	(51.17)	(50.06)	(51.52)	(50.61)
Machine labor Cost, Rs.	15160	15008	15359	15136	15260	15072
	(8.43)	(7.19)	(8.61)	(7.36)	(8.52)	(7.28)
Total variable cost (TVC), Rs.	149356	164777	147604	163446	148480	164112
	(83.10)	(78.96)	(82.79)	(79.48)	(82.95)	(79.22)
Total cost (TFC+TVC), Rs. (US$)	179729	208688	178286	205649	179008	207169
	(2983.50)	(3478.13)	(2971.43)	(3427.48)	(2983.47)	(3452.82)

TABLE 14.1 *(Continued)*

Particulars	Villupuram district		Trichy district		Overall	
	Conventional	SSI	Conventional	SSI	Conventional	SSI
Yield (tons /ha)	104	132	101	125	102.5	128.5
Price, Rs. /ton	2300	2300	2275	2275	2288	2288
Cost of production, Rs. /ton	1421	1300	1448	1351	1434	1325
Gross returns (Rs. /ha (US$/ha))**	239200	303600	229775	284375	234469	293944
	(3986.66)	(5060.00)	(3829.58)	(4739.58)	(3907.82)	(4899.07)
Net returns (Rs. /ha)	91464	132008	83519	115558	87473	123739
	(1524.40)	(2200.13)	(1391.98)	(1925.97)	(1457.88)	(2062.32)
Benefit-cost ratio	1.62	1.77	1.57	1.68	1.60	1.73

** In the study area, 70% farmers cultivated pulses (black gram, green gram, and cowpea) as intercrops and SSI farmers would additionally generate the annual gross income of Rs. 16200/ha (270.00 US$/ha/).

Note: In this chapter, 60.00 Rs. = 1.00 US$

The planting materials were selected from the sugarcane varieties (such as Co 86032, CoV 94102 and SI 309). However, most farmers (95%) used planting materials from Co 86032 variety in both the methods. The farm machineries were widely used for plowing and ridge formation activities to reduce human drudgery.

The share of machine costs in the total cost was 8.52% and 7.28% under conventional and SSI methods, respectively. The reasons for this higher cost of inputs are:

- Although the quantity of planting materials used is less in the SSI method, yet the cost of producing single-budded chips is more. This is the reason why the cost of planting material in SSI method is at par with or marginally higher than the conventional method
- It is observed that farmers applied same quantity of fertilizers in SSI as well as conventional system.
- Human labor cost (i.e., wages) was higher in the SSI method due to more quantity of cane yield harvest.

The average yield obtained from SSI method was 128.5 ton/ha, i.e. 26% more than the yield from conventional method (102.5 t/ha). Therefore, the overall cost of production was lower in SSI (Rs. 1325/ton (22.08 US$/ton)) than THE conventional (Rs. 1434/ton (23.90 US$/ton)) method. The net returns realized from sugarcane were lower under conventional method (Rs. 87,473/ha (1457.88 US$/ha)) than under SSI method (Rs.123,739/ha (2062.32 US$/ha)). The overall benefit-cost ratio was 1.60 under conventional and 1.73 under SSI method. To sum up, the sugarcane cultivation under SSI method is more profitable than the conventional method in the study areas. However, the escalating input costs, fluctuating output prices, and delayed payments by sugar mills are the major factors limiting productivity and profitability of sugarcane cultivation.

14.3.2 CONSERVATION OF AGRICULTURAL RESOURCES

The SSI is a better method of sugarcane cultivation than the conventional methods, which are seed – water intensive. By adopting SSI method, the productivity of cane can be enhanced through drip irrigation with fertigation, maintaining optimum plant spacing and profuse tillers. The benefits of SSI method vary depending on how efficiently farmers use these practices.

One of the components of SSI method i.e. drip method of irrigation (DMI) supplies water constantly or at regular intervals at the root zone of the crops through a network of pipes with the help of emitters. Unlike furrow method of irrigation (FMI), which is widely practiced in traditional sugarcane production, the efficiency of water use is extremely high in DMI as it substantially reduces the evaporation, conveyance, and distribution losses of water (Narayanamoorthy, 1997; Narayanamoorthy, 2010; Sivanappan, 1994). Further, reports showed that the irrigation efficiency under DMI is about 90 percent whereas it was 40 percent under FMI (Dhawan, 2002; Saleth, 2009). The other benefits of SSI method over conventional method are reduced soil erosion, balanced application of fertilizers through fertigation, which enhances fertilizer use efficiency, reduced water consumption that helps to reduce over-exploitation of groundwater and electricity consumption.

The literatures indicate that the total horsepower (HP) hours of water used for drip irrigated sugarcane is about 1767 HP/ha, while on the contrary the horsepower hours used under traditional method works out to as much as 3179 HP/ha. Hence, adoption of drip method of irrigation from each acre of sugarcane can save over 44% (1412 HP-hours) of water (Narayanamoorthy, 2004; 2010). Adoption of drip irrigation system is not only reducing a substantial amount of water, but also reduces electricity due to less water consumption. The electricity consumption in SSI method was also considerably reduced by 1060 kWh/ ha over conventional method (Narayanamoorthy, 2004). Therefore, water use efficiency and electricity use efficiency are higher in SSI method than in conventional method.

14.3.3 SOURCES OF PRODUCTIVITY IN SSI OVER CONVENTIONAL METHOD

The sources of output gain in the SSI method were decomposed by estimation of Cobb-Douglas production function for SSI and conventional methods (Table 14.2). The calculated value of Chow test indicated that there was a significant difference between the sources of productivity gains under SSI and conventional methods.

The sources of productivity gains (Table 14.2) revealed that the overall contribution of difference in input-use levels to productivity gain was 20.49%, which indicated that the productivity of conventional practices can be increased by 20.49%, if the input-use levels on these farms could be increased to the levels of SSI method. In the total productivity gain due to input-use, the contribution was the highest for human labor (7.77%), followed by nitrogenous fertilizers (5.68%), micronutrients (3.53%), potassic fertilizers (2.46%), machine labor (0.54%) and phosphatic fertilizers (0.48%). Among the components of technological change, the contribution of neutral technological change in total productivity was estimated as 2.16%.

TABLE 14.2   Sources of Productivity Gain in SSI Method of Sugarcane Production

S. No.	Source of productivity difference	Contribution (%)
A.	Productivity gain due to technology change, i.e. neutral technology and non-neutral technology change	2.16
B.	Total productivity gain due to input-use	20.49
	Nitrogenous fertilizers (kg/ha)	5.68
	Phosphatic fertilizers (kg/ha)	0.48
	Potassic fertilizers (kg/ha)	2.46
	Micronutrients (kg/ha)	3.53
	Human labors (man-days/ha)	7.77
	Machine labors (hours/ha)	0.54
C.	Residual factors	1.76
D.	Total observed productivity gain	24.41

14.3.4   CONSTRAINTS FACED BY FARMERS IN SUGARCANE CULTIVATION

The problems faced in sugarcane cultivation under conventional and SSI methods were analyzed by using Garrett's ranking technique and the results are presented in (Table 14.3). The farmers expressed that the low procuring price by sugar factory was the most important problem. The second important constraint was labor shortage during critical period of sugarcane cultivation followed by rat and wild animal problems, price escalation for each operations and groundwater depletion.

In SSI technology, increase in weed growth in the wider rows is the major constraint followed by receiving subsidy for micro irrigation, clogging in drip system, non-availability of quality seed material.

TABLE 14.3 Constraints Faced by Farmers in Conventional Method of Sugarcane Cultivation

S. No	Constraints	Conventional method		SSI method	
		Mean score	Rank	Mean score	Rank
1.	Low price	70.45	I	-	-
2.	Labor shortage	69.62	II	-	-
3.	Rat and wild animal problems	58.43	III	59.67I	III
4.	Price escalation for each operations	46.58	IV	-	-
5.	Groundwater depletion	33.60	V	-	-
6.	Increase in weed growth in the wider rows	-	-	71.72	I
7.	Getting subsidy for micro irrigation	-	-	66.12	II
8.	Clogging in drip system	-	-	50.02	IV
9.	No quality seed material	-	-	42.63	V

14.4 SUMMARY

This study evaluated the profitability and resource conservation impacts of sugarcane production under Sustainable Sugarcane Initiative (SSI) method. The overall yield obtained from the SSI methods was 128.5 tons per ha, which was about 25% higher than conventional method. Although the total cost of cultivation was higher in SSI over conventional method, yet the overall gross return and net return were also higher in SSI method thereby offsetting higher costs.

It may be concluded that sugarcane cultivation under SSI method is more profitable than the conventional method. There is a significant amount of saving in irrigation water and electricity under SSI technology in sugarcane cultivation. The contribution of input use to productivity gain in SSI method was 20.49%, while technological change (2.16%) and residual factors (1.76%) accounted for meager shares. Among the inputs used, fertilizers, micro-nutrients, and human labor contributed more for productivity enhancement in SSI method.

Low procurement price, labor shortage, and wild animal problems were major constraints faced by sugarcane farmers. Therefore, extension efforts should be intensified to upscale SSI method, extending the area under drip irrigation and ensuring timely availability of critical inputs to increase the economic returns and conserve water in sugarcane production. The major policy options suggested by this study are to improve production and profitability of sugarcane include provision of drip irrigation with subsidy, necessary extension efforts, and credit support to increase the adoption of drip irrigation systems, quality supply of drip systems and service facilities to remove the clogs in the drippers and imparting periodical trainings to farmers on SSI technology.

KEYWORDS

- **resource conservation**
- **sustainable sugarcane initiative**
- **water saving**

REFERENCES

Bisaliah, S., (1977). Decomposition analysis of output change under new production technology in wheat farming: Some implications to returns on investment. *Indian Journal of Agricultural Economics, 32*(3), 193–201.

Dhawan, B. D., (2002). *Technological Change in Indian Irrigated Agriculture: A Study of Water Saving Methods* (p. 330). Commonwealth Publishers, New Delhi.

Garrett, H. E., & Woodworth, R. S., (1969). *Statistics in Psychology and Education*. Vakils, Feffer, and Simons Pvt. Ltd., Bombay, p. 329.

Gujarati, D. N., & Porter, D. C., (2014). *Basic Econometrics* (5th edn., p. 915). McGraw Hill, Boston.

ISMA (Indian Sugar Mills Association), (2015). *Indian Sugar: Statistical Abstract-Advanced Estimates*. ISMA, New Delhi, p. 2.

Mohanasundaram, P., (2015). Marketing problems faced by coir units: A study in Thanjavur District of Tamil Nadu. *International Journal of Advanced Research, 3*(4), 103–107.

Narayanamoorthy, A., (1997). Economic viability of drip irrigation: An empirical analysis from Maharashtra. *Indian Journal of Agricultural Economics, 52*(4), 728–739.

Narayanamoorthy, A., (2004). Impact assessment of drip irrigation in India: The case of sugarcane. *Development Policy Review, 22*(07), 443–462

Narayanamoorthy, A., (2010). Can drip method of irrigation be used to achieve the macro objectives of conservation agriculture? *Indian Journal of Agricultural Economics, 65*(3), 428–438.

Raju, V. T., & Rao, D. V. S., (1990). *Economics of Farm Production and Management*. Oxford and IBH Publishing Co. Pvt. Ltd., New Delhi, p. 185.

Rosegrant, W. M., & Meinzen-Dick, R. S., (1996). Water resources in the Asia-Pacific region: Managing scarcity. *Asian-Pacific Economic Literature*, *10*(2), 32–53.

Rosegrant, W. M., (1997). *Water Resources in the Twenty-First Century: Challenges and Implications for Action*. Food and agriculture, and the environment discussion paper 20, International Food Policy Research Institute, Washington D. C., U. S. A., p. 27.

Saleth, R. M., (2009). *Promoting Irrigation Demand Management in India: Potentials, Problems, and Prospects*. International Water Management Institute, Colombo, Sri Lanka, p. 152.

Shanthy, R. T., & Ramanjaneyulu, S., (2014). Socio-economic performance analysis of sugarcane cultivation under sustainable sugarcane initiative method. *Indian Research Journal of Extension Education*, *14*(3), 93–98.

Shrivastava, A. K., & Sushil, S., (2011). Sustaining sugarcane productivity under depleting water resources. *Current Science*, *101*(6), 748–754.

Sivanappan, R. K., (1994). Prospects of micro irrigation in India. *Irrigation and Drainage Systems*, *8*(1), 49–58.

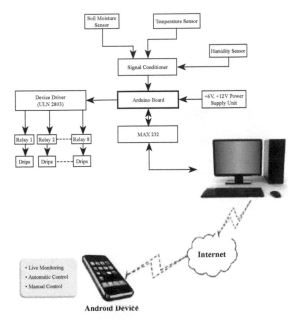

FIGURE 1.1 Overview of controls and hardwares.

	STRIP I		STRIP II		STRIP III		
	N0P0K0	N2P3K2	N0P0K0	N3P2K3	N0P0K0	N2P2K1	NPK alone
	N2P2K0	N2P1K1	N3P2K1	N2P2K3	N3P3K1	N1P1K1	
	N3P1K3 (A)	N1P1K2	N2P0K3	N1P3K1	N0P2K2 (C)	N2P3K3	B I
	N1P2K2	N3P3K2	N2P2K2	N3P3K3	N3P2K2	N2P1K2	
OUTS	N0P0K0	N3P2K3	N0P0K0	N2P2K1	N0P0K0	N2P3K2	NPK+ 6.25 t ha⁻¹ FYM
	N3P2K1	N2P2K3	N3P3K1	N1P1K1	N2P2K0	N2P1K1	
	N2P0K3 (B)	N1P2K1	N0P2K3 (C)	N2P3K3	N3P1K3 (A)	N1P1K2	B II
	N1P2K2	N3P3K3	N3P2K2	N2P1K2	N1P2K2	N3P3K2	
	N0P0K0	N2P2K1	N0P0K0	N2P3K2	N0P0K0	N3P2K3	NPK+ 12.5 t ha⁻¹ FYM
	N3P3K1	N1P1K1	N2P2K0	N2P1K1	N3P2K1	N2P2K3	
	N0P2K3 (C)	N2P3K3	N3P1K3 (A)	N1P1K2	N2P0K3 (B)	N1P2K1	B III
	N3P2K2	N2P1K2	N1P2K2	N3P3K2	N2P2K2	N3P3K3	

Treatment structure

1. N0P0K0	5. N1P1K1	9. N2P1K1	18. N3P1K1
2. N0P1K0	6. N1P1K2	10. N2P0K2	19. N3P2K1
3. N0P0K0	7. N1P2K1	11. N2P1K2	20. N3P2K2
4. N0P1K2	8. N1P3K2	12. N2P2K0	21. N3P2K3
		13. N2P2K1	22. N3P3K2
		14. N2P2K2	23. N3P3K3
		15. N2P2K3	24. N3P1K3
		16. N2P3K2	
		17. N2P3K3	

FIGURE 2.1 Layout plan of STCR –IPNS experiment with transgenic cotton under drip fertigation (Field 76, Eastern block, TNAU, CBE).

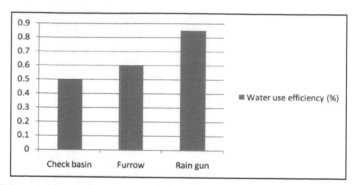

FIGURE 6.1   Water use efficiency for different irrigation methods.

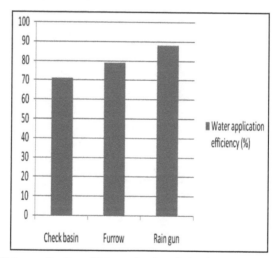

FIGURE 6.2   Water application efficiency for three irrigation methods.

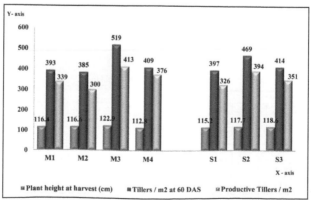

FIGURE 10.1   Influence of irrigation levels and nitrogen doses on plant height and tillers/m².

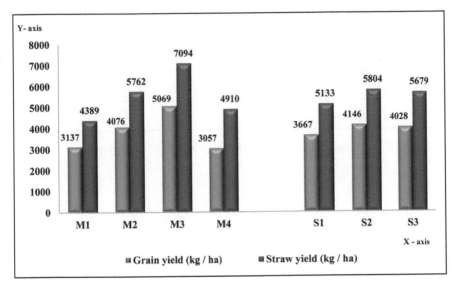

FIGURE 10.2   Grain and straw yield as influenced by levels of irrigation and fertigation.

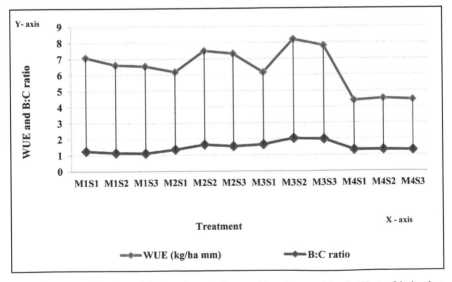

FIGURE 10.3   WUE and B:C ratio as influenced by the combined effect of irrigation levels with nitrogen doses.

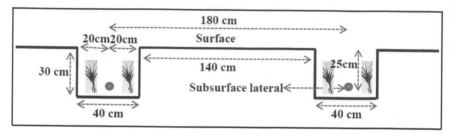

FIGURE 12.1   Design and layout subsurface drip irrigation with fertigation.

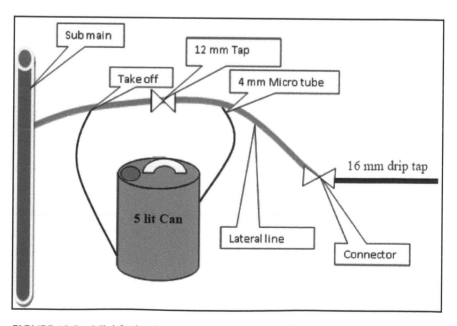

FIGURE 12.2   Mini fertigation unit for fertigating individual row.

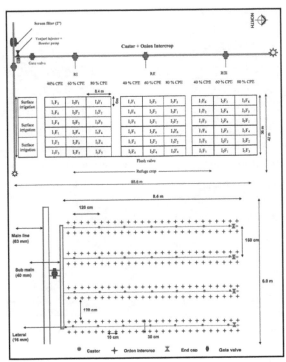

FIGURE 15.1   Experimental layout (top) and drip fertigation system (bottom): Castor + Onion intercrops.

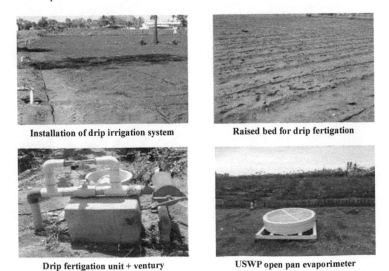

Installation of drip irrigation system   Raised bed for drip fertigation

Drip fertigation unit + ventury   USWP open pan evaporimeter

FIGURE 15.2   Components of drip irrigation system and Class A Pan.

Soil nutrient dynamics

Observation of castor yield attributes

Harvesting of castor

Experimental view of castor + onion intercrop

Onion intercrop planted on either side of castor

Onion intercropped in castor

Harvesting of onion intercrop

Marketable yield of onion intercrop

FIGURE 15.3   Onion intercrop in the castor = onion intercropping system.

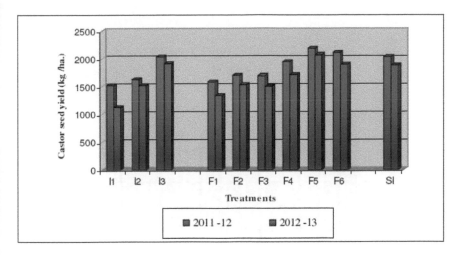

FIGURE 15.4   Castor seed yield (kg/ha) as influenced by drip fertigation, during 2011–12 and 2012–13.

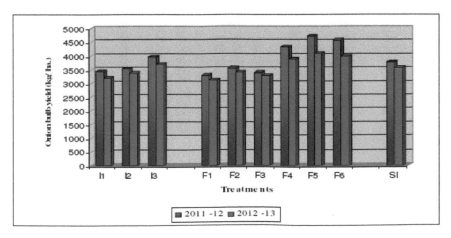

FIGURE 15.5   Marketable onion bulb yield (kg/ha) as influenced by drip fertigation: 2011–12 and 2012–13.

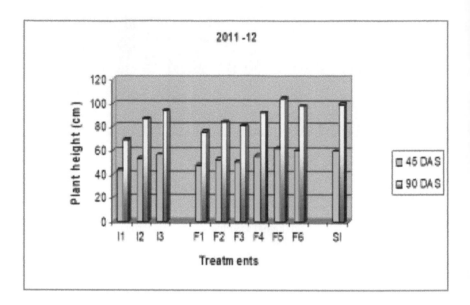

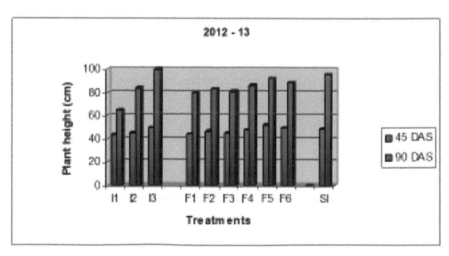

FIGURE 15.6 Effects of drip fertigation on plant height (cm) of castor: 2011–12 and 2012–13.

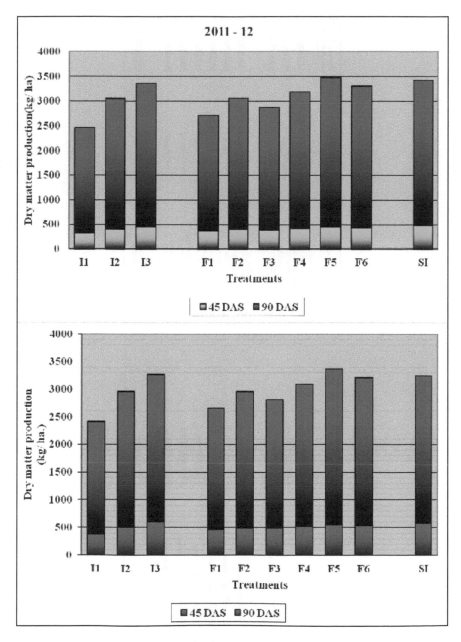

FIGURE 15.7 Effects of drip fertigation on dry matter production (DMP, kg/ha) of castor: 2011–12 and 2012–13.

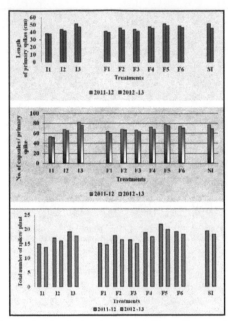

FIGURE 15.8 Effects of drip fertigation on yield attributes of castor: 2011–12 and 2012–13.

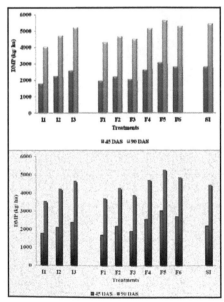

FIGURE 15.9 Effects of drip fertigation on dry matter production (kg/ha) of onion intercrop, during 2011–12 and 2012–13.

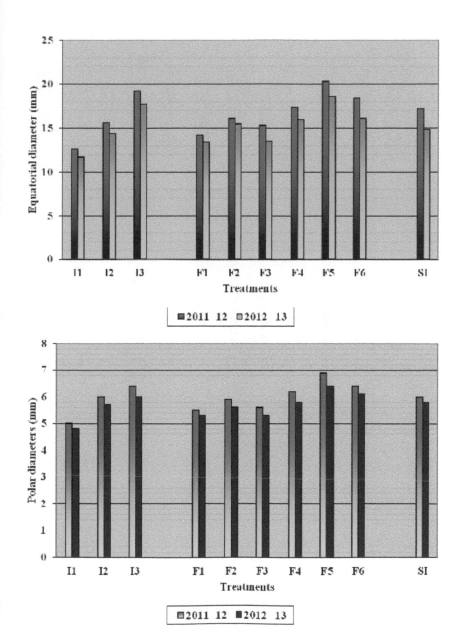

FIGURE 15.10a   Effects of drip fertigation on yield attributes of onion intercrop, during 2011–12.

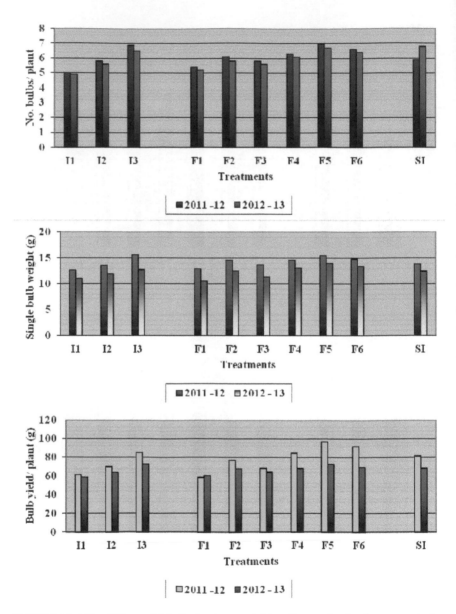

FIGURE 15.10b Effects of drip fertigation on yield attributes of onion intercrop, during 2012–13.

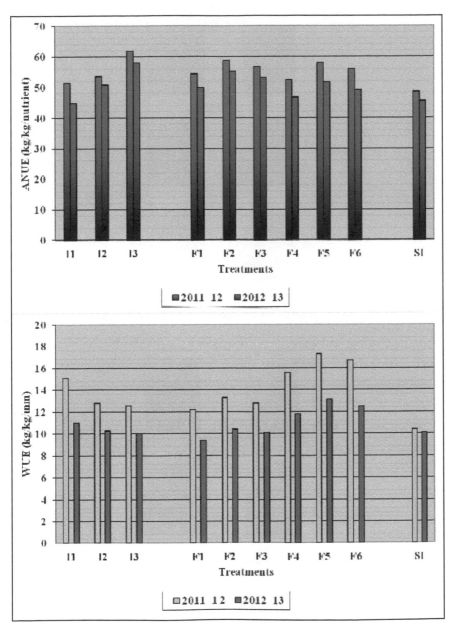

FIGURE 15.11 Effects of drip fertigation on agronomic efficiency and water use efficiency of castor + onion intercrop: 2011–12 and 2012–13.

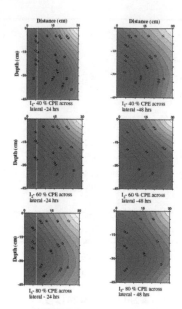

FIGURE 15.12   Soil moisture (% weight basis) distributions across drip lateral for castor + onion intercrop.

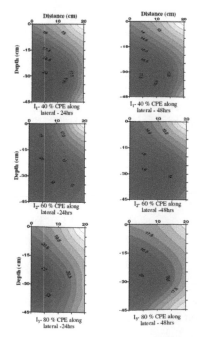

FIGURE 15.13   Soil moisture (% weight basis) distributions along the drip lateral, for castor + onion intercrop.

A. Before transplanting of cauliflower

B. After transplanting and imposing the treatment

FIGURE 17.1   Cabbage vegetable crop in the field.

Field observation

Discussion with JICA

FIGURE 17.2   Field harvesting of cabbage.

PART V
Performance of Drip Irrigated Crops

CHAPTER 15

PERFORMANCE OF A CASTOR + ONION INTERCROPPING SYSTEM UNDER DRIP FERTIGATION

K. R. PUSHPANATHAN

ABSTRACT

This chapter focuses on the effects of drip fertigation in castor + onion inter-cropping system on growth/ yield attributes and yield, water use and nutrient use efficiency, and cost economics. Drip irrigation at 80% CPE with the application of 100% RDF through WSF recorded significantly higher values of growth and yield attributes castor seed yield, nutrient uptake, an equivalent yield of castor and onion intercrops than the other treatments. Length of the primary spike, number of capsules per spike and the total number of spikes per plant of castor were significantly improved at 80% CPE with 100% RDF as WSF application. Drip fertigation at 80% CPE with 100% RDF as WSF recorded higher castor seed yield (2619 and 2490 kg/ha).

Drip irrigation at 80% CPE with 100% RDF as WSF registered higher uptake of nutrients (N, P, and K) in castor. Drip irrigation at 80% CPE with 100% RDF as WSF recorded increased plant height, more number of leaf sheaths and dry matter production of onion intercrop. The equatorial diameter, polar bulb diameter, number of bulbs plant^{-1}, single bulb weight and bulb yield plant^{-1} were significantly influenced by drip irrigation at 80% CPE and fertilizer application at100% RDF as WSF than other treatments. Drip fertigation at 80% CPE with 100% RDF as WSF significantly recorded higher marketable onion yield (5316 and 4577 kg/ha) compared to other treatment combinations.

The combined application of irrigation at 80% CPE with fertilizer at 100% RDF as WSF significantly enhanced the nutrient uptake (N, P, and K) of onion intercrop. Under drip irrigation, at 80% CPE with the application of 100% RDF through WSF significantly recorded higher castor equivalent yield (6416 and 5759 kg/ha) compared to other treatment combinations. Application of 100% RDF as WSF with 80% CPE showed higher agronomic efficiency (AE) (66.1 kg^{-1} of nutrient^{-1}). Significantly higher water use efficiency (WUE) was recorded under drip irrigation at 40% CPE with 100% RDF as WSF combination (19.3 kg/ha-mm^{-1}).

Under drip irrigation at 80% CPE maintained higher soil available moisture at 30 cm distance across the lateral at 15–30 cm depth for 24 and 48 hours after irrigation. At the end of fertigation across the lateral, the peak available nitrogen significantly higher at 100% RDF as WSF in the depth of 15–30 cm at 15 and 30 cm from the dripper. The peak availability of phosphorus recorded just below the dripper. Higher available potassium recorded near the emitter point and decreased as the distance increases with a lower level to 30 cm depth on distance and depth. The combination of drip fertigation at 80% CPE with 100% RDF as WSF significantly recorded superior land equivalent ratio (LER) value of 1.59 and 1.52 compared to other treatment combinations.

Drip fertigation at 80% CPE with 100% RDF (75% CF + 25% WSF) recorded maximum net return of Rs. 1, 63, 921 per ha (2732 US$/ha) to surface irrigation with soil application of 100% RDF as conventional fertilizer. Drip irrigation at 80% CPE with 100% RDF as WSF resulted in higher net present value (Rs. 97,019 per ha (1617 US$/ha)) and the discounted benefit-cost ratio (2.30) than other treatment combinations.

15.1   INTRODUCTION

Castor (*Ricinus communis* L.) is a native of tropical Africa (Weiss, 1971), and is also grown in Asia, Central, and North America and Europe (Doan, 2004). Among the nine cultivated oilseed crops, castor has great industrial and commercial value (Padmavathi and Raghavaiah, 2004). India, China, Brazil, Russia, Thailand, Ethiopia, and the Philippines are major castor growing countries in the world (Damodaram and Hedge, 2011). India contributes to 85% of world castor production (Weiss, 1971) and 80–90% of the world requirement of castor oil. In India, Gujarat holds the first rank with regards to the area of 0.49 million ha (57.0%), production of 0.896 million tons (82.9%) and productivity of 2010 kg/ha (2010 –11) as reported by Patel and Patel (2012).

Castor oil is being as a new source for biodiesel, and *Ricinoleic acid* is major unsaturated fatty acid constituting 87% of castor oil (Akpan et al., 2006; Baldwin and Cossar, 2009; Glaser et al., 1993; Osava, 2003). Therefore, castor crop can be grown for bioenergy purpose (Vignolo and Naughton, 1991).

In South India, castor is a drought tolerant crop plant well-adapted to low moisture conditions by deep rooting and can thrive under conserved moisture. Higher yields of castor can be realized with moderate rainfall of 600–700 mm and fairly good yields have been obtained with a well-distributed rainfall of 375–500 mm. Water management can enhance castor productivity and water use efficiency (Raj et al., 2010; Severino et al., 2012; Souza et al., 2007).

Scheduling irrigation in terms of when, how much to irrigate is important to enhance productivity both under irrigated and limited irrigated conditions. The benefits of irrigation at different stages of crop growth to increase castor yield (DOR, 1993). Subbareddy et al. (1996) reported that supplemental irrigation of 50 mm either at early (0–45 DAS) or mid (45–90 DAS) stress period gave 26% additional castor yield. The periods of vegetative, primary, and secondary spike developmental stages were critical stages of moisture. Irrigation management practices imposed at mid stress stage resulted in higher production of secondary spikes.

The castor plant is extremely sensitive to excess soil moisture at all stages of growth. Castor being deep-rooted crop can extract water from considerable soil depths. As compared to heavy soils, light textured soils require more frequent irrigations. The first irrigation is given on pre- or post-sowing irrigation for better germination; second irrigation is generally given at 3–4 leaf stage, the third at 6–8 leaf stage. The subsequent irrigations are given at 15–20 days intervals depending on soil moisture conditions. The most critical period for moisture is terminal bud initiation to the full flowering of the primary spike. Hence at this stage, adequate moisture in seedbed should be ensured by proper scheduling of irrigation.

Malavia et al. (1995) found that drip irrigation at 0.3–0.9 CPE was superior to surface irrigation. Economic feasibility of drip irrigation was assessed on castor hybrid GCH 4 grown on sandy loam soils. Drip irrigation of 0.8 fraction of pan evaporation (FPE) recorded higher seed yield (2635 kg/ha), the extra income of Rs. 2,280 per ha (380 US\$/ha), highest water saving (62.3%), water expense efficiency (8.36 kg/ha-mm^{-1}) and additional irrigated area (1.65 ha) compared to surface irrigation. Patel et al. (2006) found comparable castor yield with fertigation of a reduced dose of N in Gujarat.

As the plant growth advances, daily consumptive use increases and evapotranspiration demand decreases with crop maturity (Sabale and Khuspe, 1986). Pahalwan and Tripati (1984) indicated water use range of 455 to 600 mm at 0.6 and 0.8 IW/CPE ratio, respectively. Irrigation at 0.9 IW/CPE ratio at later stages and at 0.6 IW/CPE ratio during early growth increased the water use efficiency (Subramanian et al., 1974).

Drip fertigation increases crop yield and nutrient use efficiency; reduces the quantity of leaching of nutrients and fertilizer demand of a crop. It saves water and fertilizer and helps to maintain the desired concentration and distribution of nutrients in the soil. At Junagadh – Gujarat, drip irrigation scheduled at 0.6 CPE and fertigation of 100% nitrogen significantly increased castor seed and stalk yield with remarkable high water use efficiency and net returns as compared to surface irrigation and soil application of nitrogen (Lakkad et al., 2005). Patel et al. (2006) reported that fertigation of 100 kg of N gave significantly higher seed yield (3655 kg/ha), maximum gross return, net return and the benefit-cost ratio (BCR).

Under the cropping system approach in agriculture, complementarities arise due to better utilization of resources (below ground and above ground) both in temporal and special dimensions as in case of intercropping. The intercropping systems results in enhanced produce per unit area and time accrued due to duration differences, and rooting systems compared to sole cropping.

Being a long duration crop with slow growth habit in the initial stages castor is being grown in wider spaced rows. This feature offers a potential scope for using its interspace for growing short duration and quick growing intercrops (Singh and Singh, 1988). Intercropping is one of the potential cropping systems to use natural resources more efficiently than a single crop (Srilatha et al., 2002). Rajput and Srivastava (1996) evaluated the influence of legume intercrop components on castor and were found to increase production and net profit.

Castor and onion crops respond positively to irrigation and nutrient application with respect to growth and yield. Among the several agronomic practices, irrigation combined with fertigation are two important aspects for proper crop growth and yield. Irrigation and fertigation not only affect the growth and yield of castor + onion intercrop but also influence the quality. In South India, a detailed investigation on the response of castor to drip fertigation with onion as intercrop is lacking.

Considering all these aspects, the field studies were proposed to evaluate the effect of drip irrigation and fertigation on growth, yield, and economics of *Rabi* castor + onion crop intercropping system with the following objectives:

- To standardize the irrigation schedule for castor + onion intercropping under drip irrigation system.
- To optimize the fertigation schedule under drip fertigation system for castor.
- To monitor the soil nutrient and moisture dynamics under drip fertigation system.
- To determine the production potential, physiological variation and economics of drip fertigation under castor + onion intercropping system.

15.2   REVIEW OF LITERATURE

Castor crop is cultivated between 40°S to 52°N (Weiss, 1983) and it requires a moderately high temperature of 20°–26°C with low humidity throughout the growing season to produce maximum yield and cannot withstand frost.

15.2.1   WATER REQUIREMENT OF CROPS AND THE IMPORTANCE OF DRIP IRRIGATION

In India, castor is usually raised as a rain-fed crop because of its harsh and drought resistant characteristics. In India, the rain-fed castor crop is grown in the *Kharif* season and a partially irrigated crop in the *Rabi* season mainly in Gujarat state. Castor uses soil moisture more effectively when N, P, and K fertilizers are given (Singh and Ramakrishna, 1975). Time of irrigation is very important for castor during the primary raceme to avoid water stress. During flowering, adequate soil moisture is essential during hot climates. Shortage of moisture during this period results in a high percentage of lighter seed. Irrigation is not required during 21–28 days prior to harvest. Irrigation can also increase absorption of photosynthetically active radiation, due to an increase in total leaf area.

Wali et al. (1988) stated that castor grown during *Kharif* season under irrigated conditions produced significantly superior yield per plant, 100 seed yield and seed yield (2130 kg/ha) than the rain-fed conditions. Irrigated castor gave 57% higher seed yield than rain-fed treatment. Patel and Jaimini (1991) reported that a number of capsules per plant are positively correlated with seed yield, irrespective of the management practices.

Subbareddy et al. (1996) inferred that most sensitive stages for reducing the bean yield of castor were a vegetative stage, the formation of primary spikes and secondary spikes. Castor under stress-free moisture conditions resulted in 42% additional bean yields compared Torain-fed cultivation. Maintaining available soil moisture at low water tension and almost constant during the entire growth period through micro irrigation with considerable water saving up to 50% was possible with micro irrigation (Patel et al., 2006).

15.2.2 IMPACT OF DRIP IRRIGATION ON SOIL MOISTURE RETENTION AND CROP PHENOLOGY

Bucks and Davis (1986) have listed several benefits of using drip irrigation including an increase in yield of tomato (Bafna et al., 1993) besides the high initial cost of drip system (Nakayama and Bucks, 1991). The drip system has resulted in water saving of 40 to 80% (Locascio et al., 1989). Several researchers (Hagin and Lowengart, 1996; Martin et al. 1994; Papadopoulos, 1992; Sampathkumar et al., 2006; Segel et al., 2000; Solaimalai et al., 2005) have confirmed uniform and better crop performance under drip irrigation compared to surface irrigation.

15.2.2.1 SOIL MOISTURE DISTRIBUTION PATTERN UNDER DRIP IRRIGATION

Patil (1999) and Li et al. (2003) have observed that the movement of water in the soil depends on the soil and dripper characteristics. Bharambe et al. (2001) stated that 1.00 ETc drip with alternate day irrigation helped to maintain soil moisture content near the field capacity. However, there was always greater than 50% available water content in irrigation depths of 0.60 ETc, 0.80 ETc, and 1.00 ETc, respectively. According to Liu et al. (2003), increasing the applied volume had little effect on the horizontal wetted area.

Vishalakshi et al. (2007) reported that the pattern of soil wetting was circular and that of soil profile was elliptical under a single emitter. Under two emitters, the nature of wetting at the soil surface tended to become elliptical whereas the moisture front advance in the profile changed from a three dimensional to almost two dimensional, after the moisture front of two emitters has touched each other. The pattern of soil moisture distribution under emitter was bulb-shaped (Suganya et al., 2007). The moisture content was near the field capacity under drip irrigation (Arulkar et al., 2008). Patel et al. (2008) studied the effects of irrigation levels based CPE on the soil moisture distribution under drip irrigation.

15.2.2.2   IMPACT ON GROWTH COMPONENTS

Castor shoots: Sudhakar and Praveen Rao (1998) stated that irrigation at 0.8 IW/CPE ratio increased yield attributes and seed yield of castor compared to the lower values. Naghabhushanum and Raghavaiah (2005) inferred that plant height of castor was increased by 32.7 and 24.5% at 0.8 IW/CPE compared to 0.4, 0.6 IW/CPE ratios, respectively.

   Other crops: Praveenrao and Raikhelkar (1993) stated that irrigation levels of sesame at 0.9 ID/CPE ratio produced significantly higher dry matter per plant when compared to 0.6 or 1.2 ID/ CPE ratio. Dhurandher et al. (1995) reported that irrigation at 0.7 IW/CPE ratio resulted in maximum dry matter yield of soybean. Singh et al. (2000) reported that the sunflower plant height (157 cm and 156 cm) recorded was higher at 0.6 and 0.9 IW/CPE ratio than at 0.3 ratio. Kumavat and Dhakar (2000) revealed that irrigation at 0.8 IW/CPE ratio during critical growth stages of soybean gave significantly higher dry matter production and a number of branches per plant compared to the control.

15.2.2.2.1   Impact on Rooting Pattern

Irrigation significantly influences the soil factors and affected root growth parameters (Klepper, 1991). Michelakhis et al. (1993) reported that root density (RD) was increased significantly from water use level at 0.3 to 0.6 Ep (evaporation), and slightly at 0.6 to 0.9 Ep level and >60% of the roots were within the upper 0.5 m soil layer. According to Machado et al. (2000), most of the tomato root system was within 40 cm of the soil profile.

15.2.2.2.2   Impact on Yield Components, Yield, and Crop Quality

Castor: Naghabushanam et al. (2005) reported that irrigation level at 0.8 IW/CPE gave higher castor yield than at 0.4, 0.6 IW/CPE. Moreira et al. (2009) observed that daily drip irrigating based on 25%, 50%, 75%, 100% and 125% of Class A Pan Evaporation (ECA) influenced the weight of 100 seeds of 2nd and 3rd orders of the total racemes of castor bean.

Patel et al. (2010) observed highest castor seed yield of 2841 kg/ha under 0.8 ADFPE (Alternate day fraction pan evaporation) compared with 0.6 and 0.4 ADFPE. Raj et al. (2010) observed that length of the spike, capsules per spike, spikes per plant, seed yield per primary spike and seed yield per plant for Gujarat castor hybrids (GCH 5) were superior higher with 6 cm of irrigation at 0.8 IW /CPE. Sudharani et al. (2009) observed a 32% increase in seed yield with irrigation at 0.75 IW/CPE ratio compared to the rain-fed castor.

Other crops: Yazar et al. (2002) found the highest average corn yield of 1920 kg/ha with irrigation at 100% CPE once every six days. Lamm (2005) indicated 0.75 conjunctive use of the water use by corn under surface irrigation.

15.2.2.3   IMPACT ON WATER USE EFFICIENCY AND WATER SAVING

Castor: Ramanjaneyulu et al. (2010) reported castor water use efficiency 4.36 kg/ha-mm^{-1} with irrigation at 100 mm CPE. Lakshmamma et al. (2010) reported that genotype of castor (RG 2714) was valuable germplasm source to transfer water use efficiency to higher yielding and well-adapted cultivars.

Other crops: Micro irrigation system resulted in 30 to 70% water saving in various orchard crops and 10 to 60% increase in yield of vegetable crops compared to gravity irrigation (www.ikisan.com). Janat and Somi (2001) indicated 35% water saving in cotton compared to surface irrigation. Veeranna et al. (2000) reported 42% higher water use efficiency in chili under drip irrigation compared to furrow method. Drip irrigation water requirement of 330.46 mm for sweet corn was reported by Viswanathan et al. (2002).

Bangar and Chaudhari (2004) observed higher water use efficiency of 114% and water saving of 41% compared to values under furrow irrigation, with 125% of 200: 115: 115 kg of NPK ha⁻¹in sugarcane (cv. Co–86032) under drip irrigation. In drip-irrigated corn, Nazirbay et al. (2005) reported 371 mm of water use compared to 547 mm in furrow irrigation system. According to Wentworth and Jacobs (2006), 52% less water was used in surface micro irrigation than that under border check.

Aujla et al. (2005) observed26% increase in WUE of cotton compared to check basin. Irrigating daily at 75% showed comparable corn yield (Sorensen and Butts, 2005). According to Simsek et al. (2005), the WUE can be improved if the investigator knows when to irrigation based on crop water status.

Sampathkumar et al. (2006) observed 71% of WUE under drip irrigated cotton compared to 10% in alternate furrow irrigation. Barros et al. (2008) observed maximum dry matter production of castor (BRS −149) at 100% available soil water than at 40% available soil water level.

Patel et al. (2008) reported the highest WUE in groundnut at 40 mm of CPE. Pandey et al. (2009) observed WUE of 50.7 kg/ha-mm⁻¹with 8 lph emitter, 46.8 kg/ha-mm⁻¹ with 4 lph dripper, and 29.7 kg/ha-mm⁻¹ minisprinkler for bitter gourd. Bandyopadhyay et al. (2009) found a significant reduction in WUE of cotton increase in irrigation depth.

15.2.3 IMPACT OF MAJOR NUTRIENTS ON CASTOR: CROP GROWTH AND YIELD

According to Kittock et al. (1967), N-application in irrigated castor increased the castor yield. Thadoda et al. (1996) observed higher values of growth and yield parameters in castor GCH 4 at 100 kg N ha⁻¹. Srivastava and Chandra (2009) observed higher castor seed yield with 80 kg-N ha⁻¹ than zero nitrogen. Application of 30 kg-P ha⁻¹ produced comparable yield compared with 60 kg-P ha⁻¹. Application of 30 kg-K ha⁻¹ produced significantly taller plants than the control.

Narayan et al. (2009) observed a significant increase in plant height, number of leaves per plant, number of branches per plant, and yield of castor with a combination of RDF with poultry manure than RDF alone. Sririsha et al. (2010) observed maximum plant height and dry matter production of castor at 75% of Nitrogen from 100% of 60: 30: 30 kg NPK ha^{-1}combined with 25% N through poultry manure compared to the control.

Sarma (1985) observed that seed yield of GAUCH–1 castor hybrid was significantly higher than Aruna (13.69%) and Bhagya (24.98%). Madhu-sudana Rao and Venkateswarlu (1988) observed a significant increase in castor bean yield, when N level was increased up to 60 kg/ha. Vishnu-murthy (1988) reported that average castor bean yield was increased from 300 to 700 kg/ha by application of an optimum dose of fertilizer combination @ 50 kg-N + 30 kg-P per ha. Patel et al. (1991) obtained higher castor yield of 2,440 kg/ha, when 75 kg-N/ha was used in three splits (half as basal, half in two equal splits at 40 and 70 DAS (days after sowing).

With an increase in N up to 60 kg/ha, a number of capsules was significantly 25% more than without nitrogen (Vijayakumar, 1992). Vijaya-kumar and Shivashankar (1992) stated the bean yield of castor was at par with each other at 60 and 90 kg-N/ha $^{-1}$and was superior over the control. Mathukia and Mothwadia (1993) used 3 splits of 75 kg-N/ha (N_{75} ½ + ¼ + ¼) to obtain a 14.14% higher yield compared to N_{75} as basal. Rao et al. (1995) obtained 12% higher castor yield with 75 kg-N/ha 22 kg-P/ha compared with unfertilized plots. Thadoda et al. (1996) reported highest GCH–4 castor yield of 2130 kg/ha and stalk yield of 2470 kg/ha with 100 kg-N/ha compared with the values at 50 kg-N/ha.

Madhusudana and Venkateswarlu (1988) reported a decrease in castor oil content with the increase in N dose. They found significantly higher oil content with fertilization compared no fertilizer under rain-fed conditions. Patel et al. (2005) reported higher castor yield with two splits of 80 kg-N/ha compared to a single dose of 40 kg-N/ha. Narayan et al. (2009) observed highest castor bean yield of 2657 kg/ha with a combination of RDF with poultry manure @ 3 t/ha. Sririsha et al. (2010) observed castor bean yield of 713 kg/ha with 75% RDF + 25% N via poultry manure over the control. Patel et al. (2009) reported higher castor bean yield of 3015 kg/ha with the application of 120 kg-N/ha compared with 40 and 80 kg-N/ha.

Raj et al. (2010) observed higher oil castor bean content with cumulative pan evaporation ratio of 60 mm depth in furrow irrigation than other irrigation methods. Patel et al. (2010) observed 49% higher oil content in castor at 0.8 ADFPE (Alternate day fraction pan evaporation). However, it was at par with 0.6 ADFPE. Patel and Patel (2012) observed an increase in castor bean yield with 150: 25: 0 kg of NPK.

15.2.4   EFFECTS OF DRIP FERTIGATION ON CROPS

Fertigation is the efficient and precise application of WSFs through micro irrigation (Billsegars, 2003; Hartz and Hochmuth, 1996). Drip fertigation may: reduce the risk of crop damage in many crops due to high water table coupled with heavy rains; allow nutrients directly into root zone; increases nutrient use efficiency (Kozhushka et al., 1994).

Appropriate frequency of fertigation at weekly, bi-weekly or monthly is the best to maximize nutrient uptake by crop (Hochmuth and Smajstrla, 2000; Thompson et al., 2003). The frequent irrigation permits favorable conditions for movement of soil moisture within the wetting zone and uptake by roots (Segel et al., 2000). Generally, fertigation should be done early in the crop cycle with small amounts of nutrients, then increase the rate of application of nutrients with the crop growth rate and an increase in nutrient demand. Near crop maturity, nutrient rates can be decreased slightly (Hochmuth, 1992; Tyler and Lorenz, 1991).

According to Kovach et al. (1999), fertigation combined with daily irrigation reduces the risk of emitter clogging. Deek et al. (1997) reported a high tomato yield of 47.1 t/ha with N fertigation in ten equal splits for the equal time of intervals compared to three equal splits.

Hebbar et al. (2004) recommended the use of 100% WSF) to minimize clogging. Prabhakara et al. (2010) observed 176.7 g/plant of total DMP of chili with subsurface fertigation at 10 cm depth. Shinde et al. (2000) observed that application of 50% RDF as WSF through drip irrigation produced almost identical cotton yield compared to that of 100% RDF with the surface method of irrigation and resulted in saving of 50% fertilizer. Kumavat and Dhakar (2000) observed soybean yield of 9.40 g per plant with fertigation of 60 kg N per ha when compared to the control. Tumbare and Bhoite (2002) observed beneficial effects of concluded that weekly fertigation in 14 equal splits for green chili. Tumbare and Nikam (2004) revealed that fertigation of 100: 50: 50 kg per ha of NPK at 2-day intervals up to 105 days resulted in a significantly higher yield of green chili 9.06 t per ha.

Rani (2006) suggested that fertigation of 50% recommended N and K was the best agronomic practice to boost the yield of baby corn. Prabhakara et al. (2010) revealed that values of yield attributes were higher when compared to rest of the treatments except weekly subsurface fertigation at 10 cm depth. Drip fertigation of urea phosphate @150% RDF enhanced the productivity of maize and okra (Selvarani, 2009).

15.2.4.1   EFFECT OF DRIP FERTIGATION ON SOIL NUTRIENT DISTRIBUTION

15.2.4.1.1   Nitrogen Concentration

Bar-Yosef (1999) stated that drip fertigation with a higher dose of nitrogen (74 kg/ha) resulted in higher EC, soluble K and NO_3-N in soil compared to lower N doses @39 and 58 kg N/ha.

The available Nitrogen moved laterally up to 15 cm and vertically up to 5 cm and thereafter it was declined. The available K moved both laterally and vertically up to 15 cm and thereafter was reduced (Bangar and Chaudhari, 2004). Tumbare et al. (1999) concluded that out of 125, 100, 75, 50 and 25% of liquid fertilizer (8: 8: 8 kg/ha of NPK), fertigation of 75% dose performed better, resulting in 25% saving of fertilizer than the conventional method of placement of fertilizer.

15.2.4.1.2   Phosphorus Concentration

Potato yield was increased from 24.2 tons ha^{-1} for no P fertigation compared to 40.6 t/ha for 60 mg $liter^{-1}$ of P application via a trickle system (Papadopoulos, 1992). Satisha (1997) found that the efficiency of phosphorus fertilizer could be increased by 45% by drip fertigation compared to only 10–20% achievable by the conventional method of application. Bharambe et al. (2001) reported that the highest amount of available P was confined to the top 0–15 cm layer just immediately below the emitter.

Phosphorus has not been generally recommended for fertigation by trickle irrigation system because of its high clogging potential and limited movement in soil Singh et al. (2003) recommended not to fertigate Phosphorus. Harjinder et al. (2004) stated that 100% P applied at planting time of potato through trickle irrigation by dissolving fertilizer in water showed more available P in the entire soil depth.

15.2.4.1.3   Potassium Concentration

Bar-Yosef and Sagiv (1985) indicated that at time of maximum nutrient uptake rate by several crops grown under drip irrigation, K must be supplied through the water even when it is in sufficient concentration (as exchangeable ion) in the soil. Bharambe et al. (2001) revealed that application of RDF per plant for fresh and ratoon crop of banana in 4 splits through drip with alternate day irrigation found maximum available K in the surface soil. Further, it was observed that under high moisture regime 1.00 ETc, there was considerable movement of K up to 30 cm depth. A similar movement of K was noticed under irrigation in the control.

Suganya et al. (2007) observed almost no Potassium movement to deeper layers. Mnolawa (2008) reported that the dynamics of Potassium nitrate (KNO_3) around a dripper are influenced by the presence of plants. In maize, the soil water content, soil electrical conductivity, soil water solution and mass of KNO_3 were decreased substantially more than when there was no maize. The difference in dynamics of KNO_3 with and without maize is attributed to the uptake of KNO_3.

15.2.4.2 NUTRIENT USE EFFICIENCY OF CROPS UNDER FERTIGATION

Castor: Mathukia and Modhwadia (1995) reported higher NPK contents due to Nitrogen fertilization in castor. The castor with an average yield of 1700 kg/ha removed 50 kgN, 20 kg-P_2O_5 and 16 kg-K_2O (Jacob and Vexkull, 1958).

Other crops: The FUE was improved in tomato (Carrijo and Hochmuth, 2000; Singandhupe et al., 2003), chili (Veeranna et al., 2000) and potatoes (Chawla and Narda, 2001), compared basal methods. Solaimalai et al. (2005) reported a reduction in the leaching of Nitrogen under fertigation compared to the traditional system.

15.2.4.2.1 Nitrogen

Bangar and Chaudhari (2004) suggested the highest significant uptake of nutrients uptake in sugarcane at drip fertigation of 125% WSF on alternate days compared to 100% RDF under surface irrigation. Srivastava and Chandra (2009) found a significant increase in castor bean yield with an increase in Nitrogen dose. The crop responded to

K application up to 30 kg-K_2O per ha. Sampathkumar and Pandian (2011) observed significantly higher nutrient uptake in hybrid maize with fertigation of 150% of 225:112.5:112.5 kg NPK per ha than 100% of RDF.

Thompson et al. (2000) observed a higher amount of residual N at 0–90 cm soil depth when supplied with higher N rates and irrigated at high soil moisture tension. However, N was lost beyond root zone when irrigated at low soil moisture tension. Singandhupe et al. (2003) observed 8% higher uptake of Nitrogen in tomato with fertigation than that under furrow irrigation. Janat (2004) reported that Nitrogen recovery ranged between 48–55% in fertigated cotton compared to 43% for surface irrigated cotton. The average total N-uptake for cotton was 145 for control compared to 417 kg-N per ha for fertigation.

15.2.4.2.2 Phosphorus

Deficiency of Phosphorous at early growth stages should be avoided (Grant et al., 2001; Ryan, 2002). Tumbare and Nikam (2004) stated that application of RDF at every irrigation (2 days interval) up to 105 days recorded significantly higher uptake of Phosphorus (12.58 kg/ha) in chili than that under surface irrigation (8.53 kg/ha).

15.2.4.2.3 Potassium

Prince et al. (1998) observed that foliar-K concentration in capsicum plants was more under fertigation with 1: 1 ratio of N and K fertilizer application. Higher K-content in leaf was obtained under fertigation (Obreza and Vavrina, 1995). Anilkumar (2001) reported that K-uptake was significantly higher (105.14 kg per ha) with 100% irrigation than 75% of irrigation level on the basis of at 0.8 IW/CPE ratio.

Tumbare and Nikam (2004) stated that application of RDF at every irrigation (2 days interval) up to 105 days recorded significantly higher K-uptake (99 kg/ha) by chili than under surface irrigation (44 kg/ha). Sampathkumar and Pandian (2011) observed a higher uptake of K with 150% RDF through fertigation once in six days.

15.2.4.3 PERFORMANCE OF CASTOR CROP UNDER DRIP FERTIGATION

Lakkad et al. (2005) revealed that fertigation of 100% Nitrogen significantly increased castor bean and stalk yield compared to surface irrigation. Raj et al. (2010) reported that castor hybrid (GCH 4, 5) irrigated with furrows at 60 mm of irrigation recorded significantly higher oil yield compared to other irrigation methods, but its oil yield was at par with alternate furrow irrigation with 45 mm of irrigation (Kamalkumar et al., 2009).

Patel et al. (2010) stated that fertigation of 100% recommended dose (RD) of nitrogen resulted in higher castor bean yield compared to that at 50% RD of Nitrogen fertigation and 100% RD of Nitrogen through spot application.

15.2.4.4 WATER STRESS VERSUS CROP PARAMETERS

15.2.4.4.1 Effect of Water Stress on Relative Water Content (RWC)

Sausen and Goncalves Rosa (2010) found that leaf water potential was adversely affected due to increasing soil moisture deficits. The effects of moisture stress were emphasized by Li and Staden (1998) and Viswanathan et al. (2002) on maize; Song et al. (1995). Jiang and Huang (2000) indicated that plants suffering from water stress at tasseling stage had significantly lower RLWC values compared to those at the vegetative stage. Leaf relative water content was significantly affected by water stress in cotton (Ennahli and Earl, 2005).

Babita et al. (2010) reported that castor hybrids showed higher RWC with an increase in the osmotic adjustment with increasing stress period up to 33 days. Keyvan (2010) reported that with an increase in the intensity of water stress showed decreased relative water content of bread wheat cultivars at critical stages of crop growth.

15.2.5 INTERCROPPING SYSTEMS IN CASTOR CROP

Intercropping of castor with cluster bean (Venkateswarlyu and Reddy, 1989), with green gram (Purshotam Rao et al., 1989; Singh and Singh,

1988) has shown higher bean yield compared to sole castor. Ashokkumar et al. (1995) stated that castor as a sole crop (100 cm between rows) at wider spacing without and with intercrop proved superior to the millet-castor intercrop. Patel et al. (2007) observed higher castor bean equivalent yield (3895 kg/ha) and B:C ratio (3.41) under intercropping with green gram or sesame compared to sole castor.

Leelarani (2008) stated that highest castor yield 1265 kg/ha was recorded in castor + pigeon pea (2: 1 ratio) followed by castor + green gram (1157 kg/ha) over sole castor crop (1130 kg/ha) and it was reduced with castor + sunflower (860 kg/ha) and castor + maize (1003 kg/ha). Singh (2009) reported that castor + green gram and castor + moth bean (1: 1 ratio) produced comparable castor yield; and castor bean equivalent yield recorded with castor + green gram 1: 1 (4531 kg/ha) was 837 kg/ha higher compared to sole castor.

Patel et al. (2010) observed highest castor bean yield of 2734 kg/ha with paired row planting (180 –60 –180 cm) x 60 cm compared to normal planting (120 x 60 cm). Sharathkumar et al. (2010) indicated that castor bean yield in sole cropping system was comparable with yield obtained in paired row system of castor + cluster bean and of castor + groundnut (2: 4); whereas castor + cluster bean in 2: 4 or 1: 3 and castor + groundnut in 2: 4 row proportion recorded the higher castor bean equivalent yield of 2380, 2345 and 2136 kg/ha, respectively.

Basith and Mohammed (2010) reported that intercropping of two rows of black gram between castor with uniform row spacing 90 x 60 cm produced an additional yield of 338 kg/ha. Similar advantages were also recorded by intercropping three rows of black gram between the paired rows of castor 120 x 60 cm by obtaining an additional yield of 287 kg/ha with no loss in castor yield compared to the sole crop of castor.

Subbareddy et al. (1996) stated that castor (GAUCH–4) during *Kharif* season with 75,000 plants per hectare showed an increase in growth components (such as dry matter, leaf area index) during different crop growth stages resulting in higher Nitrogen uptake compared to 38,000 plants per hectare.

15.2.6   PERFORMANCE OF ONION UNDER DRIP IRRIGATION AND FERTIGATION

In sandy loam soils, more frequent irrigation from 11 to 20 produced onion (*Allium cepa* L.) yield of 17 to 27.4 t/ha, respectively (Singh and Sharma,

1991). According to Corgan et al. (2000), sandy soils, having the disadvantage of requiring more frequent irrigation, and also the nitrogen leaching may be more of a concern than other soils. Daily irrigation of onion in sandy loam soils increased the growth and bulb yield compared to irrigation on alternate days (Konton et al., 2003). Kumar et al. (2006) reported that though India occupies the first position in the area (0.59 m-ha), yet it occupies second place in production (7.5 m-tons) after China.

Fontes (1998) observed that there was a rapid increase in plant height from the 56th to 84th day after sowing. Halvorson et al. (2002) observed that the rate of total dry matter production of onion was maximum at 60–75 days after planting.

15.2.6.1 WATER STRESS VERSUS ONION PERFORMANCE

Begum et al. (1990) reported that irrigation at 0.40 bar soil water potential gave higher yield (12.57 t/ha) than at 0.60 bar (6.6 t/ha). Rajput and Neelam (2006) found that 56.4 cm of irrigation water resulted in the highest yield of onion under micro irrigation. Abdullah et al. (2005) observed that soil moisture deficit affects at all growth stages of onion.

15.2.6.2 MICRO IRRIGATION VERSUS ONION PERFORMANCE

Suitability of drip irrigation over furrow irrigation for winter vegetables (like onion, radish, carrot, spinach, tomato, turnip, and cauliflower) was studied by Clark and Smajstrla (1995) and they reported an increase in yield of these winter vegetables with drip irrigation. Onion crop responds well to drip irrigation (Pandita, 2000; Segel et al., 2000). Patil et al. (2000) recorded higher onion bulb yield with 53.69% water saving using drip irrigation.

Bhonde et al. (2003) observed that bulb weight was higher (1050 g) by 23.53% in drip compared to flood method (850 g) and the bulb yield per ha was increased by 13.76% under drip over flood irrigation. Hanson et al. (2003) found that onion yield was significantly lower when irrigating only once a week. According to Patil (2005), drip irrigation at 100% ET resulted in higher marketable onion yield (46.7 t/ha) with an increase of 11% than drip irrigation at 75% ET (39.5 t/ha) and 31.5% when compared to surface irrigation.

15.2.6.3   WUE OF ONION

Drip irrigation with 50% ET recorded highest WUE (1380.4 kg/ha cm^{-1}) followed by 75% and 100% ET treatments, which recorded 1252.7 kg/ha cm^{-1} and 1111.6 kg/ha cm^{-1}, respectively (NRCOG, 2002). Blaine (2003) found that onion crop irrigated with 60, 75, 90, 110 and 120% ET gave WUE of 40.9, 37.4, 41.3, 43.4 and 46.8 kg/ha-mm^{-1}, respectively. Konton et al. (2004) reported that micro irrigation resulted in water saving of 32% over furrow irrigation.

15.2.6.4   FERTIGATION VERSUS PERFORMANCE OF ONION

Hartz and Hochmuth (1996) stated that drip fertigation reduced overall fertilizer application rate and can minimize adverse environmental effects. Due to the increase in N-fertigation level, plant growth and bulb parameters were increased (Patil et al., 2000). Singh and Verma (2001) indicated 25% saving with N-fertigation. Dawelbeit and Richter (2004) obtained onion yield of 29t/ha with alternate day fertigation. Rajput and Neelam (2006) recorded the lowest onion yield with fertigation monthly. Savitha et al. (2010) reported that fertigation with 75% RDF (75: 112.5: 56.25 kg of NPK per ha) registered higher bulb onion yield of 12.70 t/ha compared to soil application of fertilizer.

15.2.6.5   FERTIGATION VERSUS FERTILIZER USE EFFICIENCY

Fujiyama and Nagal (1987) opined that nutrient solution along with irrigation water was a superior method of saving nutrients and water. Bharambe et al. (1997) found that N fertilizer use efficiency (FUE) was considerably increased with nitrogen fertigation @ 75 kg of N ha^{-1} over soil application of fertilizer. Renault and Wallender (2000) stated that onion needs high irrigation frequency and better irrigation scheduling to increase fertilizer use efficiency due to a reduction in leaching of N.

According to Ruiz and Romero (1998), NPK promotes plant development and increase the uptake of nutrients by the plants. Balasubramanyam et al. (1999) estimated that about 18 t/acre of onion bulbs will remove about 70: 25: 55 kg of NPK. The 60 kg-N/ha caused a significant improvement in bulb yield (Dimri and Singh, 2005). The performance of onion was significantly better with 200 kg-N/ha (Reddy and Reddy, 2005).

All growth parameters of onion showed a linear relationship with the increase in P dose. The uptake of P was declined beyond 50 kg/ha (Anjani-kumar et al., 2000). The K deficiency can cause reduced onion yield (Singh and Verma, 2001). Al-Moshileh (2001) reported that Potassium @ 150 kg/ha reduced the unmarketable bulbs yield compared to the control. Onion crop takes K nearly equivalent quantity to N (Salo et al., 2002). Yadav et al. (2003) concluded that the 150 kg-K/ha gave 10.01 and 30.57% more onion bulb yield compared to K @ 100 and 150 kg/ha. The 50, 100 and 150 kg-K/ha resulted in 6.11, 14.85 and 9.47% more onion yield compared to control, respectively (Kumar et al., 2006).

15.2.6.6 EFFECT OF FERTIGATION ON SOIL MOISTURE AVAILABILITY

Howell et al. (1981) stated that the wetting soil was a bulb symmetric pattern, which was two dimensional, under drip irrigation system. Singh et al. (1990) reported that the moisture distribution was more uniform within 10 cm radius of the emitter with maximum uniformity at zero, while no uniformity was noticed when distance from the emitter was increased, and also the waterfront advanced rapidly in the beginning and the rate of advance was decreased with time (Mishra and Pyasi, 1993). In Coimbatore, surface irrigation showed a steep decline of available soil moisture from 90 to 24% whereas available soil moisture was consistent under drip irrigation system throughout the irrigation cycle (once in 2 days) at about 87% and it was always nearer to the field capacity (Bobade, 1999).

15.2.7 FERTIGATION VERSUS CASTOR INTERCROPPING SYSTEM

Castor being a long duration and widely spaced crop, it offers a great scope for using its interspace for growing short duration intercrops (Singh and Singh, 1988). Intercropping in castor increased the production and net profit per unit area per unit time (Rajput and Mishra, 1995). Intercropping is one of the potential cropping systems to use natural resources more efficiently than a single crop (Srilatha et al., 2001).

Castor and cluster bean intercropping (1: 2 ratio) produced higher LAI and DMP compared to sole crop at different growth stages (Subbareddy et al., 2004). Sharathkumar et al. (2010) stated that higher castor equivalent yield of 2380 kg/ha was obtained in castor + cluster bean intercropping in 2: 4-row proportions; and the land equivalent ratio was higher (1.71) compared to other intercropping and sole cropping systems. Subbareddy et al. (2004) reported that intercropping of castor and cluster bean recorded higher castor bean equivalent yield (31 and 40%) at 30–45 cm soil depths compared to sole crops. They also stated that intercropping of castor + cluster bean (1: 2) recorded higher gross returns by 30 and 52% compared to the sole crop of castor (7,111 Rs./ha (119 US$/ha)) and cluster bean (6,080 Rs/ha (101 US$/ha)). Cauliflower – hybrid vegetable crop sequence with a drip at a low level of irrigation (IW/CPE of 0.5) gave higher yield and saved irrigation water compared to other crop sequences with furrow irrigation (Ashokkumar and Singh, 2006). Sharathkumar et al. (2010) observed significantly higher castor bean yield (1434 kg/ha) in sole cropping than the yield under intercropping systems except for paired row systems of castor intercropped with cluster bean and groundnut.

15.2.8   INTERCROPPING VERSUS LAND EQUIVALENT RATIO

Pooran Chand and Sujatha (2000) reported that castor intercropped with a black gram in 1: 6 ratio row proportion resulted in higher land equivalent ratio (LER) value of 1.85 coupled with a higher yield of castor. Srinivasa Rao et al. (2006) reported that pigeon pea + sorghum intercropping recorded higher LER value of 1.31 compared to other intercropping systems. Mohamed Amanullah et al. (2006) reported that cassava intercropped with cowpea registered higher LER than cassava intercropped with maize. Venkateswarlu (1986) revealed that two and three rows of black gram grown as intercrops between two lines of pigeonpea at 90 x 120 cm row spacing recorded higher LER (1.67 to 1.68) compared to other intercropping systems with green gram and cowpea. Itnal et al. (1994) reported that LER was highest (1.41) in intercropping of pearl millet + pigeonpea in row proportion of 4: 2 followed by 3: 1-row proportion (1.36). Omprakash and Bushan (2000) found that intercropping treatments showed higher values of LER than sole cropping treatments. Pigeonpea/castor +greengram intercropping showed the highest LER (1.62 and 1.61). Significantly higher values of LER also indicated better utilization of castor + legume combinations compared to intercropping with sesame or sorghum.

15.2.9 ECONOMICS OF DRIP FERTIGATED CROPS

Drip irrigation system reduces weed infestation, pest occurrence and enhances water and nitrogen use efficiency. Higher net income per unit water consumption and additional net income over the conventional method of irrigation for either on a hectare basis or on equal water usage is more promising. Although initial investment and maintenance costs are high, yet returns are equally high compared to traditional irrigation methods. Gala (1992) reported that the net gain in returns under drip irrigation system was considerably higher (30–60%) than the surface irrigation. Sivanappan (1998) revealed an extra income of Rs. 49,280 (821 US$) under drip irrigation in tomato over surface irrigation and the pay-back period of drip system was only six months. Drip fertigation with 100% recommended NPK registered highest BCR in white onion (Balasubramanyam et al., 2001) and in chili (Tumbare and Bhoite, 2002).

Patel et al. (2006) observed a higher gross return, net return, and BCR under 100% N-fertigation compared to 80% N-fertigation of castor. Bangar and Chaudhari (2004) reported that net extra income due to fertigation with WSF in sugarcane was 6.57% higher (59,190 Rs./ha (987 US$/ha)) than fertigation with straight fertilizer (N as urea through fertigation, P, and K as basal). Sharathkumar et al. (2010) reported that castor + cluster bean recorded higher net return of 37,938 Rs./ha (632 US$/ha) and B: C ratio of 3.36 followed by castor + cluster bean and castor + groundnut and comparable bean yield with sole castor crop.

The right choice of intercrop in wider spacing would efficiently utilize the available resources for maximizing the production, additional income from crop, long run fertility management, and environmental sustainability.

15.3 MATERIALS AND METHODS

15.3.1 EXPERIMENTAL SITE

During *Rabi* season of 2011–12 and 2012–13 and to study effects of drip irrigation and fertigation levels on growth and yield of castor + onion intercropping system: Field experiments were carried out in farmers' field (survey number 66/1) at Kokalai village of Tamil Nadu State. The soil type had 18.76% field capacity, 7.74% permanent wilting point and bulk density of 1.04 Mg m^{-3}, pH value of 8.49, the organic carbon content of 0.23% and EC 0.12 dSm^{-1}. The available status of N, P, and K in the initial soil sample was low. These soil parameters were determined according to the methods described by authors listed below:

- Field capacity (%): Pressure plate apparatus (Richard, 1947).
- Permanent wilting point (%): Pressure membrane apparatus (Richard, 1947).
- *In situ* Bulk density (g cc^{-1}): Dakshinamurthy and Gupta (1968).
- Infiltration rate (mm h^{-1}): Dakshinamurthy and Gupta (1968).
- Maximum water holding capacity (%): Dakshinamurthy and Gupta (1968).
- Textural composition (%): International Pipet method (Piper, 1966).
- Textural class: Sandy loam, USDA – SCS Classification.
- Soil pH: Glass electrode (Jackson, 1973).

The first and second crops of castor + onion intercrop were sown during *Rabi,* 2011–12 (Mid Oct. – Apr.) and *Rabi,* 2012–13 (Oct. to Mar.), respectively. The castor hybrid (YRCH 1) seeds used for the study were procured from Castor Research Station in Tamil Nadu. The onion variety CO 3 used for the study was procured from Tamil Nadu Agricultural University, Coimbatore.

15.3.2   EXPERIMENTAL DESIGN AND LAYOUT

The treatment layout plan of the experiment is given in Figure 15.1. The treatments comprised of three drip irrigation regimes and six drip fertigation levels and were compared with surface irrigation. The experimental site was irrigated by open well water. Drip irrigation system was operated once in 3 days on the basis of cumulative pan evaporation (CPE);, and water was applied as per the treatment schedule. Fertigation was carried out once on alternate drip irrigation. Fertigation was based on fertigation schedule prepared for castor + onion intercrop system. The split-plot randomized design was used. Under surface irrigation as control, water was applied based on IW/CPE ratio of 0.75 as suggested by Sudharani

(2000). The depth of irrigation water was fixed as 5 cm. The treatments consisted of irrigation and fertilizer levels as follows:

Main plots – three drip irrigation regimes were:	Sub plots – six fertigation levels were:
I_1: 40% CPE	F_1: 75% RDF through conventional fertilizer (CF)
I_2: 60% CPE	F_2: 75% RDF through water soluble fertilizer (WSF)
I_3: 80% CPE	F_3: 75% RDF (75% CF + 25% WSF)
	F_4: 100% RDF through conventional fertilizer (CF)
	F_5: 100% RDF through water soluble fertilizer (WSF)
	F_6: 100% RDF (75% CF + 25% WSF).

Control: Surface irrigation at 0.75 IW/CPE with soil application of 100% RDF through conventional fertilizer (CF) for comparison.

Where:

CPE - Cumulative Pan Evaporation

RDF - Recommended dose of fertilizer

CF - Conventional fertilizer

WSF - Water soluble fertilizer

Design: Split-Plot Design

Replications: Three

RDF, kg/ha: 60 of N + 30 of P_2O_5 + 30 of K_2O

15.3.4 FERTILIZER APPLICATION

The sources of NPK fertilizers for fertigation through drip system was urea (46% N), Muriate of Potash (60% K_2O) and water-soluble fertilizer poly feed (19: 19: 19 of NPK kg/ha), respectively. Table 15.1 indicates fertilization schedule for the castor – onion intercrops.

Conventional fertilizer treatments: Initially full dose of P (30 kg/ha) was supplied as basal dose in the form of single super phosphate (16% P_2O_5); N and K were supplied in the form of urea and Muriate of Potash, respectively as drip fertigation through venturi at 19 days after sowing on alternate drip irrigation up to 125 DAS (days after sowing).

Water soluble fertilizer treatments: The fertilizers were dissolved in water separately (for 1 kg urea, 1 liter of water; 1 kg poly feed of 19:19:19 with 2 liters of water were used for dissolving) and then they were mixed in a container and fertigated through venturi as per the fertigation schedule is given in Table 15.1. Fertigation was started from 19 days after sowing once on alternate drip irrigation up to 125 DAS. During the process of fertigation, initially the system was run for 15 minutes without fertilizers after the irrigation, conventional fertilizer treatment and water-soluble fertilizer treatments were drip fertigated through venturi, and then alone irrigation water was given (Middle rule of fertigation) to allow cleaning of lines. Irrigation was regulated by lateral valves fitted near to the take-off points of the sub main (Figure 15.1).

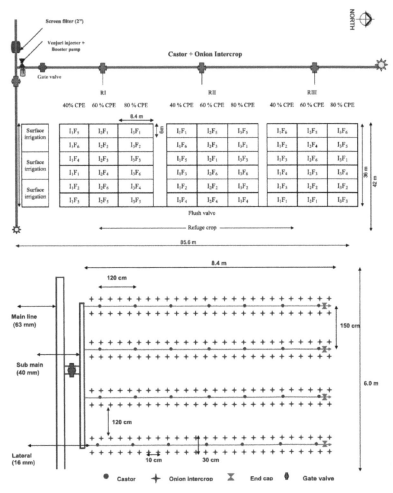

FIGURE 15.1   (See color insert.) Experimental layout (top) and drip fertigation system (bottom): Castor + Onion intercrops.

TABLE 15.1 Fertigation Schedule for Castor + Onion Intercrops (RDF: 60: 30: 30 kg/ha NPK)

Growth phase	Duration (DAS)	Day of fertilizer application	Number of application (splits)	Conventional fertilizer		Water soluble fertilizer		
				Total quantity applied (kg)		Total quantity applied (kg)		
				N	K	N	P	K
Seedling N – 20% P – 20% K – 20%	0–20	19 DAS	1	12	6	12	6	6
Vegetative N – 50% P – 50% K – 50%	21–45	25, 31, 37 and 43 DAS	4	30	15	30	15	15
Reproductive N – 20% P – 20% K – 20%	45–100	49, 55, 61, 67, 73, 79 and 97 DAS	7	12	6	12	6	6
Maturity N – 10% P – 10% K – 10%	101–180	109 and 121 DAS	2	6	3	6	3	3
Total				60	30	60	30	30

Note: Full dose of Phosphorus (30 kg P_2O_5 per ha) was applied as basal for conventional fertilizer under drip irrigation system.

15.3.5 FERTILIZER APPLICATION IN SURFACE IRRIGATION

Fertilizer sources used for supplying NPK were urea (46% N), single superphosphate (16% P_2O_5) and Muriate of Potash (60% K_2O), of which, a full dose of P_2O_5 and 50% N and 50% of K_2O were applied as basal dose. The remaining 50% of both N and K_2O was top dressed in two equal splits at 30 and 60 DAS of the crop.

15.3.6   DRIP IRRIGATION SYSTEM (FIGURE 15.1)

Irrigation water was pumped through 7.5 HP motor and it was conveyed to the main line after filtering through screen filter. From the source line, water was taken to the field through the main line of 63 mm PVC pipes. In the main line, venturi was installed for fertigation before the screen filter. From the main line, three separate submains of 40 mm PVC pipes with control values were taken for all drip irrigation and fertigation treatments. In each submain, the laterals of 12 cm LDPE fitted with lateral control valves were taken to all the sub-plots for imposing both drip irrigation and fertigation treatments. In each plot of 6.0 m x 8.4 m bed size, four laterals consisting of inline emitters with a discharge rate of 4 lph were installed at a spacing of 60 cm. There were six subplots in each replication, each subplot consisted of four 1.5 m raised beds with four laterals running parallel to each other and the layout is shown in Figure 15.2.

FIGURE 15.2   (See color insert.) Components of drip irrigation system and Class A Pan.

Submains and laterals were closed with an end cap. After installation, the trail run was conducted to assess the mean emitter discharge and uniformity coefficient. This was taken into account for fixing the time of irrigation water application. During the irrigation period, an average uniformity coefficient of 90 to 95% was maintained. Characteristics drip irrigation systems are as follows:

Length of mainline (63 mm OD PVC)	: 90 m
Length of each sub main (40 mm OD PVC)	: 216 m
Length of each lateral from submain (16 mm LDPE)	: 1840 m
Number of laterals from submain	: 216.
Number of emitters from lateral	: 3024.
Lateral spacing	: 60 cm
Emitter type	: Inline Dripper
Emitter discharge rate	: 4 lph
Filter size (Screen filter)	: 63 mm

The drip irrigation unit was installed in the experimental site measuring 8.4 m in length and 6 m in width as per treatments. The unit consisted of a four-stage filter system with a mesh of 100 and 80 microns, water meter, control valve and air-exhaust valve attached in series to the main PVC line of 50 mm diameter. The submain pipe of 40 mm diameter with laterals of 12.5 mm diameter and point source adjustable emitters connected to the laterals were other components of the drip unit. One emitter was placed for every plant of castor crop with 1.5 m x 1.2 m spacing. For onion intercrop, two rows planted with a spacing of 30 cm x 10 cm on either side of one lateral connected with emitters were utilized.

15.3.7   CROP MANAGEMENT

The experimental field was plowed with a MB plow and harrowed thrice to bring the soil to a fine tilth. The raised beds were formed for drip fertigation plots; and ridges and furrows were formed for surface irrigation plots at 1.5 m x 1.2 m. Plots were laid out in advance according to field layout in Figure 15.1. Irrigation channels were made in the control plot for applying irrigation water. Seeds of good quality selected castor hybrid (YRCH 1) and onion bulb (CO 3) were used for sowing. The castor + onion intercrop was sown on 20.10.2011 during *Rabi* season of 2011–12 and 01.10.2012 during *Rabi* season in 2012–13.

The row spacing was at 1.5 m distance and 1.2 m between plants for castor crop. Two seeds per hill were dibbled to a depth of 3 to 4 cm by maintaining a spacing of 1.2 m between two hills in a row. In *Rabi* season of 2011–12 and *Rabi* season of 2012–13, dibbling was done in the dripper portion of one side as per field layout. Thinning was carried out 7 days after germination by allowing the healthy one and removing unhealthy plants to maintain the recommended plant population per hectare. Intercrop onion bulbs were pressed at one cm soil depth after irrigation. Table 15.2 summarizes the details for the date of sowing and harvesting of castor + onion intercrops.

TABLE 15.2   Details on Date of Sowing and Harvest of Castor and Onion Crops

Details	Rabi season	
	2011–12	**2012–13**
Castor hybrid, Yethapur 1		
Spacing (Drip, surface irrigation)	1.5 m x 1.2 m	
Gross plot area	8.4 m x 6.0 m	
Net plot size	6.0 m x 3.0 m	
Number of rows (gross plot)	4	
Number of rows (net plot)	2	
Date of sowing	20.10.2011	01.10.2012
Date of primary spike harvest	17.01.2012	28.12.2012
Date of secondary spike harvest	02.03.2012	11.02.2013
Date of tertiary and quaternary spike harvest	16.04.2012	29.03.2013
Total duration (days)	180	
Intercrop: Small onion variety, CO 3		
Spacing	30 cm x 10 cm	
Planting pattern	Paired row planting	
Gross plot area	8.4 m x 6.0 m	
Net plot size	5.4 m x 3.6 m	
Date of sowing, day/month/year	20.10.2011	01.10.2012
Date of harvest	17.01.2012	29.12.2012
Total duration of onion (days)	90	

All treatments were uniformly irrigated immediately after sowing. In both seasons, subsequent irrigations were given based on the cumulative pan evaporation as per the treatment. The crops were irrigated up to 150 days and thereafter the irrigation was stopped. For *surface irrigation treatment*, irrigation was given immediately after sowing followed by life irrigation on the third day through furrow irrigation to the depth of 5 cm. The concept of irrigation scheduling in surface irrigation treatments was based on IW/CPE ratio 0.75 (*i.e.,* cumulative pan evaporation value reached 66.7 mm during 2011–12 and 2012–13). Daily pan evaporation rate was recorded from the standard USWB Class A Open Pan Evaporimeter (Figure 15.2) installed in the field itself. Irrigation was given when CPE value reached 66.7 mm. The quantity of irrigation applied was according to the procedure by Sudharani (2000).

The drip irrigation treatments were imposed fifteen days after sowing of castor. Drip irrigation was applied once in three days based on CPE in *Rabi* 2011–12 and 2012–13 as per the treatment schedule for hybrid castor crop at 40, 60 and 80% of evapotranspiration. For drip fertigation system, the operating pressure was maintained at 2.0 kg cm^{-2}. Water requirement was computed as follows:

$$WRc = CPE \times Kp \times Kc \times Wp \times A \qquad (1)$$

where, WRc = Computed water requirement (liter per plant); CPE = Cumulative pan evaporation for three days (mm); Kp = Pan factor (0.8); Kc = Crop coefficient; Wp = Wetting percentage in fraction; and A = Area per plant. Crop coefficients (Kc) for castor crop (FAO, 1977) were:

Initial stage, 0–25 days	0.35
Development stage, 26–60 days	1.15
Mid-stage, 61–130 days	1.15
Final stage, 131–180, days	0.55

Duration of operation of drip system to deliver the required volume of water per plant was calculated as follows:

$$\text{Irrigation duration} = [\text{Volume of water needed}]/ \\ [\text{Emitter discharge} \times \text{No. of emitters}] \qquad (2)$$

Cumulative pan evaporation for three days and average per day was used to work out the irrigation scheduling. Accordingly, irrigation equivalent to 40% CPE, 60% CPE and 80% CPE was the irrigation level. Whenever the rainfall occurred, the drip system was not operated and 70% of effective rainfall was assumed to be equal to the quantity of water to be applied through the drip (Yellamandareddy and Sankarareddy, 2005). The computed water requirement (WRc) for each irrigation treatment was calculated (Eq. (2)) and irrigation water was supplied once in three days using the emitter of 4 lph. The amount of water to be applied through each emitter under different levels was randomly measured based on pan evaporation.

The weeds present between and within the crop rows were removed manually at 30 DAS and subsequent weeding was done as and when required. The crop was well protected when semi loopers became a problem due to heavy rainfall during October by spraying Monocrotophos @ 0.05% and from botrytis disease by spraying carbendazim @ 0.05%.

The castor crop was harvested in 3–4 pickings based on the maturity of the main spikes and the spikes that are formed on secondaries, tertiaries, and quaternaries. The harvested spikes were sun-dried and threshed by manual beating with sticks. The threshed produce was winnowed and seeds were cleaned. The seed and stalk yield for each plot was recorded separately after drying.

15.3.8   GROWTH PARAMETERS OF CASTOR

Five plants from the net plot area were selected at random and tagged. The tagged plants were used for recording all biometric observations at different growth stages of the crop. The number of days to 50% flowering in the treatment were observed and calculated. Five plants were randomly selected within the net plot area and were tagged to record the plant height at 45 DAS and 90 DAS. This was recorded in the field (non-destructive sampling) and was measured on the main stem from the base of the plant (ground surface) to the tip of the apical bud. The total number of leaves in each of the five selected plants was counted and the average was expressed at 45, 90 and 135 DAS for calculating the leaf area index of the crop. The sampled plants at 45 and 90 DAS were collected and dried in hot air oven at $65+ 5°$ for four to five days. The oven-dried samples were weighed, and total dry matter production was calculated and expressed in kg/ha. The leaf area index, crop growth rate, specific leaf weight, and relative water content were calculated following the procedure is given below:

Five castor plants were sampled at random from each plot at different growth stages for computing leaf area index (LAI). The leaf length and breadth (cm) were measured and multiplied with a crop factor of 0.516 as suggested by Patil et al. (1989) for calculating leaf area. Then the formula suggested by Williams (1946) was employed for the computation of LAI. Observations were recorded at 45, 90 and 135 DAS. Leaf area index (LAI) was calculated as follows:

$$LAI = [\text{Total leaf area plant}^{-1}\ (cm^2)]/$$
$$[\text{Ground area occupied plant}^{-1}\ (cm^2)] \qquad (3)$$

The crop growth rate (CGR, g m^{-2} day^{-1}) was computed at 45 and 90 DAS by adopting the formula by Watson (1958).

$$CGR = [W_2 - W_1]/[p\ (t_2 - t_1)] \qquad (4)$$

Where, W_1 and W_2 are dry weights of whole plants at t_1 and t_2, respectively; t_1 and t_2 are a time in days; p is spacing used (1.5 m x 1.2 m).

The specific leaf weight (SLW, mg cm^{-2}) is the leaf thickness that was determined by the formula suggested by Radford (1967).

$$SLW = [\text{Leaf dry weight (mg)}]/[\text{Leaf area (cm}^2)] \qquad (5)$$

Relative water content (RWC, %) in fully expanded 3rd to the 5th leaf of castor crop was determined by the procedure by Barrs and Weatherly (1962).

$$RWC, \% = \{[\text{Fresh weight} - \text{Dry weight}]/$$
$$[\text{Turgid weight} - \text{Dry weight}]\} \times 100 \qquad (6)$$

Samples were collected at 45, 90 and 135 days after sowing for analysis. To determine the plant RWC, twenty-five leaf discs were collected in each treatment and weighed (fresh weight, FW) immediately in the fully expanded leaf of the plant. The weighed leaf discs were then placed in distilled water Petri-discs for 4 hours at 25°C and then the turgid weight (TW) was measured. The samples were then dried in a hot air oven at 72°C for 48 hours to obtain the dry weights (DW). Then the RWC was calculated.

15.3.9 YIELD AND YIELD CONTRIBUTING PARAMETERS OF CASTOR

Five plants from the net plot area were selected at random and tagged. The tagged plants were used for recording all yield attributes of castor crop. The yield attributes *viz.*, length of primary spike per plant, number of capsules per plant, total number of spikes per plant, 100 seed weight, biological yield, and harvest index methods followed for recording observation is mentioned below. Five tagged plants were measured from the base of primary spike emergence up to the tip of spike and expressed in cm at 90 DAS. Several capsules plant^{-1} of five randomly selected plants from each plot was noted at each picking and then the total number of capsules per plant was calculated. Five tagged plants in net plot area were counted for a number of spikes per plant at the primary spike stage, secondary spike stage, and tertiary spike stage and expressed as a total number of spikes per plant. A hundred seeds from each net plot were counted and weighed by a digital weighing machine to determine the 100 seed weight and expressed as a gram. It was calculated from the net plot area and computed on per hectare basis (kg/ha). Two central rows of each plot were harvested after maturity and put on the ground for drying. After complete drying, the plants were weighed, and biological yield was calculated and converted into ha^{-1}. *Harvest Index* is the ratio of economic yield (kg/ha) to the seed yield (kg/ha).

15.3.10   GROWTH PARAMETERS OF ONION

Five plants in each treatment were randomly selected, tagged, and the observation was recorded. The height of the plant was measured from the base to the tip of the longest leaf. The plant height was recorded on 45 DAS and at maturity (90 DAS) in five tagged plants and the mean value was calculated at each stage and expressed in cm. The total number of leaves in each of the five selected plants on 45 and 90 DAS was counted and the average was expressed at each stage of the crop growth. The sampled plants at 45 DAS and at maturity (90 DAS) were dried in hot air oven at $65 + 5°C$ till it reached a constant weight. The total dry matter production was calculated and expressed in kg/ha.

15.3.11   YIELD AND YIELD ATTRIBUTING PARAMETERS OF ONION

The equatorial diameter at the maximum circumference of the bulb (at the broadest point) was measured at harvest from five randomly selected bulbs in each of the treatment using digital Vernier caliper and the mean value was expressed in cm. The polar diameter of the bulb was measured at harvest by using digital Vernier caliper measured at the maximum girth from five randomly selected bulbs in each treatment and the mean was expressed in cm. Five tagged plants were uprooted at maturity stage (90 DAS) and counted and the average was expressed as a number of bulbs plant^{-1}. Randomly selected five bulbs from five tagged plants were cleaned and weighed in the electronic balance and the average was expressed as single bulb weight in gram. Five tagged plants were uprooted at maturity stage (90 DAS) dried for five days and cleaned leaf sheath were recorded, weighed, and the average was expressed as bulb yield plant^{-1}.

Harvesting of bulbs was done when the plants showed yellowing and drying of leaves and neck fall symptoms (Figure 15.3). The harvested bulbs along with foliage were cured for three days in the field itself, shade drying was done for two days and then the foliage was removed. The bulbs were cleaned, weighed, and expressed in kg/ha. The total bulb fresh weight was measured for each plot. The bulbs were graded to remove unmarketable bulbs (diseased bulbs, bulbs under 3.8 cm in diameter, split bulbs, double bulbs). The unmarketable bulbs per plot were weighed. The weight of unmarketable bulb per plot was subtracted from the total weight of bulbs to record the marketable bulb weight per plot. The % of marketable bulb yield was calculated as follows:

$$\text{Marketable bulb yield (\%)} = 100 \times \{(\text{Marketable bulb weight per plot}) \div (\text{Total bulb weight per plot})\} \quad (7)$$

15.3.12 CASTOR EQUIVALENT YIELD FOR THE INTERCROPPING SYSTEM

The castor equivalent yield of castor + onion intercrops was estimated as below:

$$\text{Castor equivalent yield (kg/ha)} = \{[(Yi \times Pi)/(Pc)] + [\text{Castor yield (kg/ha)}]\} \quad (8)$$

where, Yi = Yield of intercrop (kg/ha); Pi = Price of intercrop (Rs./kg); Pc = Price of castor (Rs./kg).

Soil nutrient dynamics

Observation of castor yield attributes

Harvesting of castor

Experimental view of castor + onion intercrop

Onion intercrop planted on either side of castor

Onion intercropped in castor

Harvesting of onion intercrop

Marketable yield of onion intercrop

FIGURE 15.3   (See color insert.) Onion intercrop in the castor = onion intercropping system.

15.3.13 PLANT NUTRIENT STATUS OF CASTOR AND ONION INTERCROP

The plant's samples of castor and onion were collected for estimating dry matter production at 45 and 90 DAS to oven dry at 65 + 5°C for nutrient analysis. The dried samples were ground in a Willey Mill to pass through 40 mm mesh sieve and composite sample (leaf, stem, and reproductive parts) was used for estimating the uptake of N, P, K contents of castor and onion intercrop as per standard procedures such as: Micro Kjeldahl method for N (Humphries, 1956); Triple acid digestion method for P (Jackson, 1973); and Triple acid digestion method for K (Jackson, 1973). Uptake of N, P, and K (kg/ha) was calculated based on the nutrient content in plant and dry weight of the plant. Available N, P, and K in soil were also determined. The uptake of N, P and K (kg/ha)) by castor and onion intercrop was computed as follows:

$$\text{Uptake of nutrient} = [1/100] \times [\text{Nutrient content in\% } \times \text{Total dry matter in kg/ha}] \qquad (9)$$

15.3.14 ESTIMATION OF AGRONOMIC EFFICIENCY

The agronomic efficiency of castor + onion intercrop was worked out by using the method mentioned below. Agronomic efficiency (AE, kg of yield/ kg of nutrient) was estimated by using the formula by Ravi et al. (2007):

$$\text{Agronomic efficiency} = [\text{Crop yield in kg/ha}] / [\text{Nutrient used in kg/ha}] \qquad (10)$$

15.3.15 WATER USE EFFICIENCY (WUE)

The crop water requirement, water use, effective rainfall, and water use efficiency were determined as follows:

The crop water requirement (WR, mm) was calculated from the accumulated value of water use:

$$\text{WR} = [(\text{Total water use in mm}) + (\text{Effective rainfall in mm})] \qquad (11)$$

The quantity of water applied through drip or surface system in each irrigation of different treatment was summed up to estimate total irrigation water applied (in mm).

Total water use =
[Number of irrigations x depth of irrigation in mm]   (12)

For computing total water use, the effective rainfall was also included and expressed in mm. Effective rainfall was estimated by soil moisture balance method (Yellamandareddy and Sankarareddy, 2005) and it was used to compute the total water use (in mm). Then, water use efficiency (WUE, kg/(ha-cm)) was determined as follows:

WUE = [(Economic yield in kg per ha)/(Total water use in cm)]   (13)

15.3.16   SOIL MOISTURE DISTRIBUTION

Soil moisture content was estimated by gravimetric method through soil samples taken at a radial distance (horizontal) of 0, 15, 30 cm (across lateral) and a vertical distance of 10, 20 cm (along lateral) at a depth of 0–15, 15–30 and 30–45 cm from the dripper for studying soil moisture distribution pattern in each irrigation regime. Data were collected before the start of irrigation, 24 hours and 48 hours after irrigation on non-rainy days (drip irrigation cycles) from two plants in each treatment at 90 and 135 DAS and the mean values were expressed in % soil moisture on a weight basis. Soil samples were estimated gravimetrically and collected samples were oven dried at 105°C for attaining a constant weight. The soil moisture content (%) was determined as follows:

Soil moisture content = 100 ×
{[Fresh weight, g – Dry Weight, g)/(Dry Weight, g]}   (14)

15.3.17   ROOT DISTRIBUTION

The root studies were made by measuring the tap root (rooting depth), and root dry weight (g) at crop maturity stage in both castor and onion crops and the mean values were determined. Root growth was measured using a modified trench method (Bohm et al., 1977). Trenches to a convenient depth and sufficient length on both sides of the sampling row were opened by digging with a fork without snatching the rootlets and separated away from sampling rows. The plants were carefully excavated until the tip of each plant root was just visible. The soil adhering to the root was carefully removed by immersing in a water-tub and then observations were made. Rooting depths was measured from the collar region to the tip of the deepest root and expressed in cm. Root samples were air dried initially followed by oven drying at 65 + 5°C till a constant weight was attained; and the root weight was expressed in g per plant.

15.3.18   SOIL NUTRIENT DISTRIBUTION UNDER DRIP FERTIGATION

The soil samples were taken at a radial distance of 0, 15, 30 cm (across to lateral) and 10 and 20 cm between drippers (along lateral) at a depth of 0–15, 15–30 and 30–45. Soil nutrient dynamics was estimated by analyzing the soil samples for available N, P, and K. The observations were taken between 75 and 125 days before fertigation. The mean values were calculated for both the years of experimentation and expressed in kg of nutrient per ha.

15.3.19   LAND EQUIVALENT RATIO (LER)

Land equivalent ratio (LER) is defined as the relative land area under the sole crop that is required to produce the yields obtained in the intercropping system (Willey, 1979). The LER was determined as follows:

$$LER = L_A + L_B = [Y_A/S_A] + [Y_B/S_B] \tag{15}$$

where, L_A and L_B are LER for the individual crops; Y_A and Y_B are the individual crop yields in the intercropping system; and S_A and S_B are the sole crop yields.

Sole crop of castor and onion were cultivated separately to calculate the land equivalent ratio.

15.3.20   ECONOMICS OF COST OF CULTIVATION OF INTERCROPS

The economic aspects on the cost of cultivation, gross return, net return, benefit-cost ratio, net present value, discounted benefit-cost ratio working methods for castor + onion intercrop are presented in this section. The expenditure incurred from field preparation to harvest was worked out. The total cost of cultivation was the sum of the cost of inputs and the cost of production. The total income for onion intercrop was computed using marketable onion yield and the prevailing minimum market rate of Rs. 30 per kg. Net return was worked out by subtracting total cost of cultivation from the gross returns. The benefit-cost ratio (BCR) was worked out by using the formula suggested by Palaniappan (1985).

BCR = [Gross return in Rs./ha]/[Total cost of cultivation in Rs./ha]   (16)

The cost of drip system for one hectare was worked based on current market rates. The life of the drip system was assumed to be 10 years. Prevailing market price of drip components from a standard firm was used. Interest on capital investment was taken as 8% per annum. To assess the economics of a drip irrigation system, the following items were considered for computation:

Net present value (NPV) according to Palanisami et al. (2002):

$$NPV = \sum_{i=1}^{n} \frac{(Bn - Cn)}{(1+r)^n} \qquad (17)$$

where, Bn = Benefit in the nth year; C_n = Cost in the nth year; n = Number of years; r = Discounted rate.

Discounted benefit cost ratio:

BCR = [Total discounted benefit]/[Total discounted cost]   (18)

The discounted benefit was arrived by multiplying the net additional income by the corresponding factor $(1/\{(1+i)^n\})$ for each year. It was summed up for all the ten years to arrive at the total discount benefits. Because of longer life period, the discounted benefit-cost analysis was employed to have a real-time cost-benefit appraisal of the drip system.

15.3.21 STATISTICAL ANALYSIS

The data on various parameters studied during the course of the investigation were statistically analyzed by applying the technique of analysis of variance and regression analysis that were suggested by Panse and Sukhatme (1978). Fisher's method of analysis of variance was applied for the analysis and interpretation of data. The level of significance used in 'F' and 't' tests was p = 0.05. Critical difference values were calculated, wherever the 'F' test was found significant.

15.4 RESULTS

15.4.1 CASTOR MAIN CROP

15.4.1.1 GROWTH COMPONENTS

The number of days to 50% flowering: The data revealed that the number of days to 50% flowering of castor was significantly influenced by drip irrigation regimes and fertilizer levels both during *Rabi* 2011–12 and *Rabi* 2012–13. Drip irrigation scheduled at 80% cumulative pan evaporation (CPE) recorded significantly more number of days to 50% flowering (52 days during *Rabi* 2011–12 and 48 days during *Rabi* 2012–13, respectively) than 60% CPE irrigation regime. Days to 50% flowering were earlier at 40% CPE (48 days and 47 days, respectively) during both the years.

Among the different fertilizer levels, application of 100% RDF) as water-soluble fertilizer (WSF) recorded delay in 50% flowering (53 days and 52 days) compared to 100% RDF (75% WSF +25% CF) and 100% RDF with conventional fertilizer. The number of days to 50% flowering was earlier at 75% RDF as CF in both years. Surface irrigation based on IW/CPE ratio 0.75 was able to attain53 and 50 days to 50%flowering compared to drip irrigation at 80% CPE with 100% RDF as conventional fertilizer (CF), respectively in both years.

With regard to interaction effects, the drip irrigation and fertilizer levels were able to significantly influence the number of days to 50% flowering. Treatment combination of 80% CPE with 100% RDF as WSF significantly delayed the days to 50% flowering (57 and 54 days during *Rabi* 2011–12 and *Rabi* 2012–13, respectively) than other treatment combinations.

Plant height: Castor plant height at 45 DAS and primary spike maturity stage at 90 DAS showed significant variation due to different irrigation regimes and fertigation levels during *Rabi* 2011–12 and *Rabi* 2012–13. Drip irrigation at 80% CPE at 45 DAS and 90 DAS recorded taller plants (56.8 cm, 93.2 cm during 2011–12; and 49.5 cm, 99.6 cm during 2012–13, respectively) followed by drip irrigation at 60% CPE during both years. Whereas, drip irrigation at 40% CPE recorded shorter plant height at 45 and 90 DAS (43.6 cm, 68.8 cm during 2011–12, and 43.1, 64.6 cm during 2012–13, respectively). At 45 and 90 DAS of observation, drip fertigated with 100% RDF as WSF registered higher plant height (61.3, 103.6 cm during 2011–12 and 51.8 cm, 92.1 cm during 2012–13) followed by the drip fertigation at 100% RDF (75% CF + 25% WSF). Drip fertigation with 75% RDF as CF registered lower plant height at 45 and 90 DAS (47.2, 75.7 cm during 2011–12 and 44.1 cm, 79.5 cm during 2012–13, respectively). Drip fertigation with 80% CPE at 45 DAS and 90 DAS recorded increased plant height than the surface irrigation with soil application of 100% RDF as CF (59.7, 98.7 cm during 2011–12 and 48.9, 95.8 cm during 2012–13, respectively).

The interaction effect between drip irrigation with fertigation levels was significant on castor plant height at 90 DAS. Drip fertigation at 80% CPE with 100% RDF as WSF resulted in maximum plant height at 90 DAS of 119.9 cm during 2011–12 and 112.7 cm during 2012–13 than other treatment combinations.

Dry matter production (kg/ha): The dry matter production (DMP) was influenced by irrigation regimes and fertigation levels at 45 DAS and 90 DAS during both years. During 2011–12 and 2012–13, drip irrigation regime at 80% CPE significantly recorded higher DMP at 45 and 90 DAS (454, 2905 kg/ha and 597, 2667 kg/ha, respectively) followed by 60% CPE during both years. Lower DMP was recorded at 40% CPE at 45 and 90 DAS (316, 2144 kg/ha and 374, 2040 kg/ha, respectively) in both years. Drip fertigation with 100% RDF as WSF at 45 and 90 DAS performed better and produced higher DMP of 455 and 3022 kg/ha in 2011–12 and 554 and 2818 kg/hain2012–13, respectively, and was significantly followed by drip fertigation with 100% RDF (75% CF + 25% WSF) and 100% RDF as CF in both years. The least DMP at 45 and 90 DAS (369 kg/ha, 2344 kg/ha in 2011–12 and 464, 2190 kg/ha in 2012–13) was recorded under drip fertigation at 75% RDF as CF. Generally, lower DMP of castor was recorded under surface irrigation with soil application at 100% RDF as CF (At 45 and 90 DAS: 468,2962kg/ha in 2011–12 and 573, 2676 kg/ha in 2012–13, respectively) when compared to drip fertigation at 80% CPE with 100% RDF as conventional fertilizer during both years.

The interaction effect of irrigation regimes and fertilizer levels on DMP at 45 and 90 DAS varied significantly during both years of the experiments. Among the different combinations, application of 100% RDF as WSF through drip irrigation scheduled at 80% CPE recorded significantly higher dry matter production (533,3446kg/ha at 45 and 90 DAS during 2011–12 and 648, 3170 kg/ha during 2012–13, respectively).

Leaf area index (LAI) for castor: Leaf area index (LAI) of castor at all growth stages was varied significantly for different drip irrigation regimes and fertigation levels. Drip irrigation at 80% CPE recorded significantly higher LAI values of 0.236, 0.606 and 1.206 at 45, 90 and 135 DAS during 2011–12 and 0.208, 0.585 and 1.051 at 45, 90 and 135 DAS during 2012–13 and these were at par with 60% CPE at 45 DAS during both years and significant at 90 DAS and 135 DAS in both years, respectively. The least LAI was registered at 40% CPE (0.144, 0.390, 0.716 during 2011–12 and 0.126, 0.327, 0.557 during 2012–13) at 45, 90 and 135 DAS. Among different fertilizer levels, application of 100% RDF as WSF registered higher LAI values of 0.294, 0.815 and 1.706 during 2011–12 and 0.260, 0.856 and 1.162 during 2012–13 followed by 100% RDF (75% CF + 25% WSF) and 100% RDF as CF during both years, respectively. Lower LAI value was recorded at 45, 90 and 135 DAS (0.168, 0.424 and 0.836 during 2011–12 and 0.148, 0.388 and 0.685 during 2012–13) with fertilizer at 75% RDF as CF. The lower LAI of 0.269, 0.698, 1.137 during 2011–12 and 0.232, 0.649, 1.046 during 2012–13 was recorded under surface irrigation with soil application of 100% RDF as CF compared to drip irrigation at 80% CPE with 100% RDF as CF, respectively.

The interaction effect on LAI was nonsignificant at 45 DAS in both years. Under drip irrigation at 80% as CPE with 100% RDF as WSF, the effect was significant at 90 and 135 DAS on LAI values (1.026, 2.085 during 2011–12 and 1.165, 1.390 during 2012–13, respectively) than other treatment combinations.

Crop growth rate: Significant variation was observed in crop growth rate (CGR, g m^{-2} day^{-1}) due to irrigation regimes and fertilizer levels at 45–90 DAS during both years. Among the irrigation regimes, significantly higher CGR value was observed under drip irrigation at 80% CPE (5.45 and 4.60 g m^{-2} day^{-1} during2011–12 and 2012–13, respectively) followed by drip irrigation at 60% CPE during both years. The lower CGR value was noticed in drip irrigation at 40% CPE (4.06 and 3.70 g m^{-2} day^{-1}, during 2011–12 and 2012–13, respectively) in both years. The CGR value showed variation in different levels of fertilizer application. Application of 100% RDF as WSF recorded higher CGR of 5.70 and 5.03 g m^{-2} day^{-1} during 2011–12 and 2012–13 followed by 100% RDF (75% CF + 25% WSF) and 100% RDF as CF, respectively. Lower CGR was recorded at 75% of CF in both years. Surface irrigation with soil application of 100% RDF as CF resulted in lower CGR (5.52 and 4.81 g m^{-2} day^{-1}, during 2011–12 and 2012–13, respectively) than drip irrigation at 80% CPE with 100% RDF through CF during both years.

Among different fertigation levels, drip irrigation at 80% CPE with 100% RDF as WSF recorded higher CGR at 45 and 90 DAS (6.47 and 5.61 g m^{-2} day^{-1}) than other treatment combinations during both years.

Specific leaf weight: Specific leaf weight (SLW, mg cm^{-2}) of castor under water stress and nutrient levels was significant at different growth stages of castor during both years. Higher specific leaf weight of 9.7, 9.8 and 9.3 mg cm^{-2} recorded at deficit water level (40% CPE) for 45 DAS, 90 DAS and 135 DAS followed by 60% CPE during 2011–12 and 2012–13. The lower specific leaf weight value of 7.78, 8.21 and 7.35 mg cm^{-2} was recorded at 80% CPE during 2011–12. The same trend was also recorded in 2012–13.

Lower level of nutrients (75% RDF as CF) significantly produced higher specific leaf weight (8.80, 9.10 and 8.41 mg cm^{-2} during 2011–12 and 8.79, 8.53 and 7.52 mg cm^{-2} during 2012–13, respectively) followed by 75% RDF (75% CF + 25% WSF) at different growth stages of castor during both years. Lower specific leaf weight was noticed at different growth stages in 100% RDF as WSF in both years.

With regard to interaction effect at 45, 90 and 135 DAS, maximum specific leaf weight was registered at 40% CPE with 75% RDF as CF (9.97, 9.95, 9.62 mg cm^{-2} during 2011–12 and 9.88, 10.41, 9.19 mg cm^{-2} during 2012–13) compared to other treatment combinations.

Relative water content (RWC): Relative water content (RWC,%) of castor leaves showed significant differences at vegetative stages (45 DAS), primary spike stage (90 DAS) and at tertiary spike stage (135 DAS) during both years. The percentage of RWC was lower at vegetative stage and was gradually increased at the primary spike stage and then was declined at tertiary spike stage of castor. Among drip irrigation regimes, 80% CPE resulted in higher RWC % at 45, 90 and 135th growth stages (80.1, 82.2, 79.7% during 2011–12 and 75.4, 80.0, 78.3% during 2012–13, respectively) compared to 60% CPE during both years. The minimum water content was observed at 40% CPE (70.2, 75.5, 72.3%during 2011–12 and 65.7, 71.5, 69.8%during 2012–13) at different growth stages during both years. Irrespective of different fertilizer levels, fertilizer application at 100% RDF as WSF showed higher RWC at different growth stages (80.7, 82.8, 80.9% during 2011–12 and 78.2, 81.5, 82.9% during 2012–13, respectively) followed by 100% RDF (75% CF + 25% WSF) and 100% RDF as CF during both years. Lower RWC of castor was recorded at 75% RDF as CF in both years. Surface irrigation with soil application of 100% RDF as CF recorded lower RWC than drip fertigation at 80% CPE with 100% RDF as CF in both years.

With regards to interaction effects, drip fertigation at 80% CPE with 100% RDF as WSF combination at 45, 90 and 135 DAS (84.1, 85.3, 84.6% in 2011–12 and 82.8, 85.4, 87.8% in 2012–13) significantly recorded higher RWC during both years.

Length of primary spikes: The length of the primary spike (cm) was significantly influenced under drip irrigation regimes and fertilizer levels during both years. In both years of experimentation, drip irrigation at 80% CPE significantly increased the length of the primary spike of 50.9 cm and 46.9 cm during 2011–12 and 2012–13 followed by 60% CPE. Lower length of the primary spike (38.1 in 2011–12 and 37.9 cm in 2012–13) was observed at 40% CPE during both years. Fertilizer application at 100% RDF as WSF registered maximum primary spike length (51.2 and 49.1 cm, during 2011–12 and 2012–13) followed by 100% RDF (75% CF + 25% WSF) and 100% RDF as CF. Values at the application of 100% RDF (75% CF + 25% WSF) and 100% RDF as CF were found at par with each other in primary spike length during both years. The lower spike length of 41.1 and 39.0 cm was registered in 75% RDF as CF during 2011–12 and 2012–13, respectively. Surface irrigation with soil application at 100% RDF resulted in minimum spike length of 51.3 and 45.6 cm during both years compared to 100% RDF as CF through drip fertigation.

Interaction effect on length of the primary spike was significant during both years. Treatment combination of drip fertigation at 80% CPE with 100% RDF as WSF gave maximum spike length of 57.5 and 54.7 cm per plant in both years compared to other treatment combinations.

Number of capsules per primary spike: The number of capsules per primary spike was influenced by drip irrigation regimes and fertilizer levels during both years. A higher number of capsules of 82.1 and 76.2 were produced at 80% CPE during 2011–12 and 2012–13 followed by 60% CPE during both years, respectively. The minimum number of capsules per primary spike (52.7 and 51.3) was observed at 40% CPE in both the years. Among fertilizer levels, application of 100% RDF as WSF registered maximum number of capsules per primary spike (77.1 and 75.3 during 2011–12 and 2012–13) followed by 100% RDF (75% CF + 25% WSF) and 100% RDF as CF. Application of 100% RDF (75% CF + 25% WSF) and 100% RDF as CF resulted in values comparable with each other during both years, respectively. The lower number of capsules per primary spike (63.7 and 59.6) registered at 75% RDF as CF in both years. Surface irrigation with soil application at 100% RDF recorded a minimum number of capsules per primary spike (76.9 and 69.1 in both years) compared to 100% RDF as CF through drip fertigation.

The significant interactional effect was found in a number of capsules per primary spike in both years. Drip fertigation at 80% CPE with 100% RDF as WSF combination registered maximum number of capsules per primary spike of 94.5 and 86.8 in both years, respectively.

The weight of 100 seeds (g per 100 seeds): Differences in 100 seed weight under irrigation regimes and fertilizer levels were significant during 2011–12 and 2012–13. Among different irrigation regimes, drip irrigation at 80% CPE recorded maximum 100 seed weight of 31.6 g (2011–12) and 28.3 g (2012–13) over 60% CPE. Minimum seed weight of 26.4 g and 26.1 g recorded at 40% CPE in both years, respectively. Fertilizer application at 100% RDF as WSF significantly recorded higher 100 seed weight (31.0 in 2011–12 and 28.8 g in 2012–13) than other fertilizer levels during both years. Fertilizer level at 100% RDF (75% CF + 25% WSF) and 100% RDF as CF produced comparable 100 seed weight during both years, respectively. Lesser 100 seed weight (28.3 in 2011–12 and 26.3 g in 2012–13) registered at 75% RDF as CF during both years. Surface irrigation with soil application of 100% RDF as CF registered lower 100 seed weight of 31.6 in 2011–12 and 27.8 g in 2012–13 compared to drip fertigation at 80% CPE with 100% RDF as CF, respectively. The interaction effect on drip fertigation was non-significant during both years.

Castor seed yield: Castor seed yield (kg/ha) was significantly influenced by irrigation regimes and fertigation levels during both years

(Figure 15.4). Drip irrigation at 80% CPE recorded significantly higher seed yield of 2037 kg/ha during 2011–12 and 1918 kg/ha during 2012–13, respectively followed by 60% CPE. The yield increase observed under 80% CPE was 24.5% 2011–12 and 25.0% in 2012–13 than that at 60% CPE. Lower castor seed yield was recorded at 40% CPE (1526 in 2011–12 and 1134 kg/ha in 2012–13) during both years. Among different fertilizer levels, drip fertigation at 100% RDF as WSF registered maximum seed yield of castor (2197 and 2077 kg/ha during 2011–12 and 2012–13, respectively) followed by 100% RDF through 75% CF + 25% WSF and 100% RDF as CF during both years. The percentage yield increase under drip fertigation at 100% RDF as WSF was 12.9% in 2011–12 and 20.7% in 2012–13compared to drip fertigation at 100% RDF as CF during both years. The 75% RDF as CF resulted in lower castor seed yield during both years. Surface irrigation with soil application of 100% RDF as CF gave lower castor seed yield (2042 in 2011–12 and 1881 kg/ha in 2012–13) compared to drip fertigation at 80% CPE with 100% RDF as CF during both years.

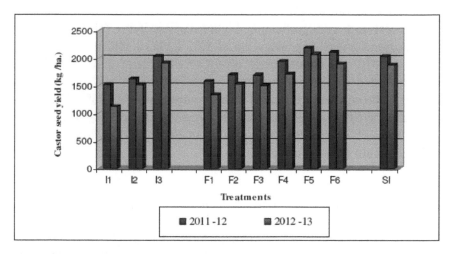

FIGURE 15.4 (See color insert.) Castor seed yield (kg/ha) as influenced by drip fertigation, during 2011–12 and 2012–13.

A significant interaction was found under different irrigation regimes and fertilizer levels during both years. Drip fertigation at 80% CPE with 100% RDF as WSF produced maximum seed yield of castor (2619 and in

2011–12 2490 kg/ha in 2012–13) compared to other treatment combinations. Irrigation regime at 80% CPE gave significantly higher pooled seed yield of castor (1978 kg/ha) compared to 60% and 40% CPE. Application of fertilizer at 100% RDF through WSF significantly recorded maximum castor seed yield in pooled data (2137 kg/ha) followed by 100% RDF through 75% CF + 25% WSF and 100% RDF as CF, whereas lower pooled castor seed yield was noticed at 75% RDF as CF.

Among the interactions, a combination of 80% CPE with 100% RDF as WSF showed higher pooled castor seed yield than other treatment combinations.

The biological yield of castor: The biological yield (kg/ha) under irrigation regimes and fertilizer levels were found significant in both years. The higher biological yield was recorded at 80% CPE (5647 and 5482 kg/ha, during 2011–12 and 2012–13, respectively) and fertilizer at 100% RDF as WSF (5712 and 5530 kg/ha, during 2011–12 and 2012–13, respectively). The lower biological yield was recorded at 75% RDF as CF in both years. The interaction effect was significantly higher at 80% CPE with 100% RDF as WSF combination (6647 and 6368 kg/ha, during 2011–12 and 2012–13, respectively).

Harvest index: Significantly higher harvest index (HI, in fraction) was recorded under drip irrigation at 80% CPE (0.360 in 2011–12 and 0.350 in 2012–13) compared to 60% CPE. Fertilizer at 100% RDF as WSF recorded maximum harvest index (0.383 in 2011–12 and 0.376 in 2012–13) during both years, and it was comparable with 75% RDF (75% CF + 25% WSF). Lower harvest index was registered at 40% CPE with 75% RDF as CF in both years. Interaction effect on drip fertigation was non-significant during both years.

15.4.1.2 NUTRIENT UPTAKE: CASTOR

The difference in nutrient uptake of castor by drip irrigation regimes and nutrient levels at the growth stages was found to be significant. Uptake of nitrogen at 45 and 90 DAS by castor was influenced by irrigation regimes and fertigation treatments during both years.

Total N-uptake: Total N-uptake (kg of N per ha) was significant under irrigation regimes and nutrient levels at 45 and 90 DAS growth stages. At 45 and 90 DAS, drip irrigation at 80% CPE significantly increased

N-uptake (18.9, 38.4 kg/ha and 16.4, 36.8 kg/ha during 2011–12 and 2012–13, respectively) compared to other irrigation regimes. For different fertilizer levels, better N-uptake was recorded at 100% RDF as WSF (20.0 and 42.0 kg/ha during 2011–12 and 16.8 and 40.0 kg/ha during 2012–13) followed by 100% RDF (75% CF +25% WSF). Lesser N-uptake of 15.9, 32.1 and 15.0, 31.4 kg of N ha^{-1} were noticed at 75% RDF as CF in both years. Surface irrigation with soil application of 100% RDF as CF showed lesser uptake of N than the drip fertigation at 80% CPE with 100% RDF as CF in a two-year study. A significant interaction was noticed under irrigation regimes and fertilizer levels. However, the interaction at 80% CPE with 100% RDF as WSF showed significantly maximum uptake of N (21.1, 46.3 and 18.1, 43.9 kg-N/ha) over other interaction treatment combinations in two years of research.

Phosphorus uptake: During 2011–12 and 2012–13, phosphorus uptake by castor (kg of P/ha) varied significantly at irrigation regimes and fertigation treatments for 45 and 90 DAS. Drip irrigation at 80% CPE recorded higher P-uptake at 45 and 90 DAS (1.20, 5.82 kg-P/ha during 2011–12 and 0.92, 5.50 kg-P/ha during 2012–13, respectively) followed by 60% CPE. Lower P-uptake by castor was recorded at 40% CPE in both years. For fertilizer treatments, application of 100% RDF as WSF recorded higher P-uptake values of 1.39 and 6.20 kg/ha during 2011–12 and 1.15 and 5.77 kg/ha during 2012–13 at 45 and 90 DAS followed by 100% RDF (75% CF + 25% WSF) as drip fertigation. Lower P-uptake was recorded at 75% RDF as CF in different growth stages during both years. Surface irrigation with soil application of 100% RDF recorded lower P-uptake at 45 and 90 DAS (1.28 and 5. 96 kg/ha during 2011–12 and 1.06 and 5.59 kg/ha during 2012–13 respectively) compared to drip fertigation at 100% RDF as conventional fertilizer with 80% CPE. The interaction effect between irrigation regimes and fertigation levels on P-uptake was significant. Significant higher interaction on P-uptake was recorded at 80% CPE with 100% RDF as WSF (1.59, 6.66 kg/ha and 1.29, 6.27 kg-P/ha) at 45 and 90 DAS in both years compared to other treatments.

Potassium uptake: Potassium uptake by castor (kg of K per ha) was significantly influenced by drip fertigation at 45 and 90 DAS growth stages during both years. Irrigation regime at 80% CPE resulted in higher K-uptake at 45 and 90 DAS (8.87 and 39.5 kg-K/ha during 2011–12 and 8.46 and 37.9 kg-K/ha during 2012–13, respectively). Lower K-uptake was recorded at 40% CPE during both years. At 45 and 90 DAS, drip fertigation at 100% RDF as WSF recorded higher K-uptake of 9.34, 41.4 kg/ha during 2011–12 and 8.86, 40.6 kg/ha during 2012–13 followed by 75% CF with 25% WSF and 100% RDF as CF in the two-year study. Drip fertigation at 75% RDF as CF recorded lower K-uptake in both years. Surface method of irrigation with soil application of 100% RDF as CF recorded lower K-uptake in different growth stages (8.71, 37.8 kg/ha during 2011–12 and 8.69, 36.8 kg/ha during 2011–12) than drip irrigation at 80% CPE with 100% RDF in both years.

Among the interactions, drip fertigation at 80% CPE along with 100% RDF as WSF had significantly recorded maximum K-uptake (9.78, 44.9 kg/ha during 2011–12 and 9.16, 43.5 kg/ha during 2012–13) during both years than other combinations.

15.4.2   ONION AS INTERCROP

The onion was intercropped on either side of castor main crop under drip fertigation to study the effects on growth and yield. Plant height of onion intercropped with castor was influenced by irrigation regimes and fertilizer levels at 45 and 90 DAS (at harvest) during both years. In general, plant height of onion intercrop was significantly higher during 2011–12 than that in 2012–13 at 45 DAS and at harvest. Drip irrigation at 80% CPE recorded maximum plant height at 45 and 90 DAS (37.9, 40.8 cm during 2011–12 and 35.6, 40.2 cm during 2012–13, respectively); whereas 40% CPE produced lower plant height during both years. For fertilizer levels, significantly maximum plant height at 45 and 90 DAS was registered at application of 100% RDF as WSF (36.5, 43.3 during 2011–12 and 38.7, 42.4 cm during 2012–13) followed by 100% RDF (75% CF + 25% WSF) and 100% RDF as CF, whereas 100% RDF (75% CF + 25% WSF) and 100% RDF as CF were comparable with each other during both years. Shorter plant height was recorded at 75% RDF as CF (32.7, 29.7 at 45 DAS and 35.2, 34.1 cm at 90 DAS, during 2011–12 and 2012–13, respectively) in both years. The interaction between different irrigation regimes and fertilizer levels were non-significant.

Number of leaf sheaths recorded at 45 and 90 DAS (harvest stage) was significantly influenced by irrigation regimes and different fertilizer levels during both years. At 45 and 90 DAS, number of leaf sheaths (21.3, 25.2 during 2011–12 and 17.5, 22.1 during 2012–13) was found to be significant at 80% CPE compared to 60% CPE; whereas lower number of leaf sheaths was noticed at 40% CPE (15.1, 19.8 and 14.1, 17.7) during both years. Among different fertilizer levels, application of 100% RDF as WSF produced more number of leaf sheaths (22.9, 27.1 and 21.0, 25.9 in both years) than other fertilizer levels. Fertilizer application at 100% RDF (75% CF + 25% WSF) and 100% RDF as CF were at par with each other at 45 and 90 DAS in both years. The lower number of leaf sheaths was observed at 75% RDF as conventional fertilizer in two years of study. Interaction effect at drip fertigation levels was non-significant during both years.

The dry matter production (DMP, kg/ha) of onion intercrop at 45 and 90 DAS was influenced by different irrigation regimes and fertigation levels. At 45 and 90 DAS, higher DMP was recorded under drip irrigation at 80% CPE (2553, 5206 kg/ha during 2011–12 and 2380, 4646 kg/ha during 2012–13, respectively) followed by 60% CPE during both years. Lower DMP was noticed at 40% CPE in both years. Among fertilizer levels, application of 100% RDF as WSF at 45 DAS and 90 DAS recorded maximum DMP value of 3048, 5620 during 2011–12 and 3018, 5252 kg/ha during 2012–13 followed by 100% RDF (75% CF + 25% WSF) and 100% RDF as CF. Application of 75% RDF as CF recorded lower DMP of 1928, 4289 kg/ha (2011–12) and 1689, 3702 kg/ha (2012–13) at 45 and 90 DAS, respectively.

The interaction effect of drip fertigation on DMP at 45 and 90 DAS were significant during both years of experiments. Combination of 80% CPE with 100% RDF as WSF significantly recorded higher DMP at 45 and 90 DAS (3449, 6220 during 2011–12 and 3330, 5653 kg/ha during 2012–13) compared to other treatment combinations.

Crop growth rate (CGR) of onion intercrop: Significant variation was observed in onion growth rate (CGR, g m^{-2} day^{-1}) due to drip irrigation regimes and fertilizer levels at 45–90 DAS during both years. Among the irrigation regimes, drip irrigation at 80% CPE significantly recorded higher CGR value (5.89 and 5.03 g m^{-2} day^{-1}) followed by drip irrigation at 60% CPE during 2011–12 and 2012–13, respectively. The lower CGR value was recorded at 40% CPE (4.99 and 3.93 g m^{-2} day^{-1}) in both years. Significantly maximum CGR value of 5.71 and 4.96 g m^{-2} day^{-1} was observed with application of 100% RDF as WSF followed by 100% RDF (75% CF + 25% WSF) and 100% RDF as CF during 2011–12 and 2012–13, respectively; whereas application of 100% RDF (75% CF

+ 25% WSF) were at par with 100% RDF as CF in both years. Lower CGR value was registered at 75% RDF as CF (5.25 and 4.47 g m^{-2} day^{-1}) in both years. Interaction effects on drip fertigation were significant on CGR during both years. Drip irrigation at 80% CPE with 100% RDF as WSF combination recorded significantly higher CGR at 45 and 90 DAS (6.16 and 5.16 g m^{-2} day^{-1}) than the other treatment combinations, during both years.

Yield contributing parameters for onion intercrop: The bulb equatorial diameter (mm) was significantly influenced by irrigation regimes and fertilizer levels during 2011–12 and 2012–13. In both years, drip irrigation at 80% CPE significantly registered higher equatorial bulb diameter (19.2 and 17.7 mm, during 2011–12 and 2012–13 respectively) followed by 60% CPE. Lower bulb equatorial diameter was recorded at 40% CPE (12.6 and 11.7 mm) in both years. Among fertilizer levels, fertigation with 100% RDF as WSF gave significantly higher equatorial bulb diameter (20.3 and 18.6 mm during 2011–12 and 2012–13, respectively) followed by 100% RDF (75% CF + 25% WSF) and 100% RDF as CF in both years. Whereas, 100% RDF (75% CF + 25% WSF) and 100% RDF as CF produced comparable equatorial diameter of the bulb in both years. Lower equatorial bulb diameter was registered at 75% RDF as CF in both years. Interaction effects for drip fertigation were non-significant in both years.

The *polar diameter of onion bulb* was significantly influenced by irrigation regimes and fertilizer levels during both years. Higher bulb polar diameter was registered at 80% CPE (6.4 and 6.0 mm during 2011–12 and 2012–13, respectively), followed by 60% CPE. The lower bulb polar diameter was recorded at 40% CPE value of 5.0 and 4.8 mm in both years. Application of 100% RDF as WSF registered higher polar diameter (6.9 and 6.4 mm) followed by 100% RDF (75% CF + 25% WSF) and 100% RDF as CF in both years, respectively. Whereas, 100% RDF (75% CF + 25% WSF) and 100% RDF as CF were at par with each other. Lower bulb polar diameter was registered at 75% RDF as CF in both years. The interaction effect was non-significant in both years.

The *number of bulbs per plant* was significantly influenced by irrigation regimes and fertilizer levels during both years. Drip irrigation at 80% CPE significantly produced more number of bulbs (6.9 during 2011–12 and 6.5 during 2012–13, respectively) followed by 60% CPE. Lower bulb number was noticed at 40% CPE (5.0 and 4.9) in a two year study. Maximum bulb number was recorded at 100% RDF as WSF (7.0 during

2011–12 and 6.7 during 2012–13) followed by 100% RDF (75% CF + 25% WSF) and 100% RDF as CF in both years; whereas 100% RDF (75% CF + 25% WSF) was found at par with 100% RDF as CF during both years. The lower number of bulbs was recorded at 75% RDF as CF (5.4 and 5.2) in both years. Interaction effects among drip fertigation treatments were non-significant during both years.

The **single bulb yield** (g per plant) was influenced by drip irrigation regimes and fertilizer levels. Significantly higher single bulb weight was recorded at 80% CPE (15.6 and 12.7g during 2011–12 and 2012–13, respectively) followed by drip irrigation at 60% CPE. Single bulb weight was declined at 40% CPE (12.6 and 11.0 g) in both years. Fertilizer application at 100% RDF as WSF gave maximum single bulb weight (15.4 during 2011–12 and 13.9 g during 2012–13) followed 'by 100% 'RDF (75% CF + 25% WSF); whereas comparable single bulb weight was observed between 100% RDF (75% CF + 25% WSF), 100% RDF as CF and 75% RDF as WSF during 2011–12. Single bulb weight was lower at 75% RDF as CF (12.9 and 10.6 g) in both years. The interaction effect was non-significant in both years.

Bulb yield per plant (g per plant) was significantly influenced by drip fertigation levels during both years. Higher bulb yield plant^{-1} was found at 80% CPE (84.9g in 2011–12 and 72.9g in 2012–13) followed by 60% CPE during both years. Lower bulb yield plant^{-1} was noticed at 40% CPE (61.4 and 58.7 g) in both years. Among different fertilizer levels, fertigation at 100% RDF as WSF had significantly recorded higher bulb yield plant^{-1} (96.6 and 72.6 g during 2011–12 and 2012–13, respectively) followed by 100% RDF (75% CF + 25% WSF) in both years. During 2012–13, values for fertilizer at 100% RDF (75% CF + 25% WSF) and 100% RDF as CF were at par with each other. Lower bulb yield plant^{-1}was recorded at 75% RDF as CF in both years. There was significant interaction under drip irrigation regimes and fertilizer levels on bulb yield plant^{-1}. Drip fertigation at 80% CPE with 100% RDF as WSF significantly produced higher bulb yield plant^{-1} (109.8 and 81.0 g) during both years.

Irrigation regimes and fertigation levels significantly influenced the **marketable onion bulb yield** (kg per ha) during both years (Figure 15.5). Marketable onion bulb yield was higher during 2011–12 than that in 2012–13. Higher marketable onion bulb yield was recorded under drip irrigation at 80% CPE (3960 kg/ha during 2011–12 and 3706 kg/ha during 2012–13, respectively) followed by 60% CPE during both years. The percentage increase in marketable onion bulb yield was 11.4 and 9.64% compared to 60% CPE during both years. Under drip irrigation at 40% CPE, lower marketable onion bulb yield was observed during both years. Among fertigation treatments, fertigation at 100% RDF as WSF produced maximum marketable onion bulb yield (4739 and 4097 kg/ha during 2011–12 and 2012–13, respectively) followed by 100% RDF (75% CF + 25% WSF) and 100% RDF as CF during both years. Lower marketable bulb yield of onion was registered at 75% RDF as CF application during both years. Surface irrigation with soil application of 100% RDF as CF recorded lower marketable onion yield compared to drip fertigation at 80% CPE with 100% RDF as CF during both years. Interaction effect of irrigation regimes and fertilizer levels was significant on marketable bulb yield production. Treatment combination of 80% CPE with the application of 100% RDF as WSF recorded higher marketable bulb yield (5316 and 4577 kg/ha during 2011–12 and 2012–13, respectively) of onion compared to other treatments.

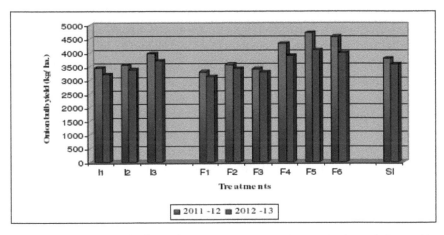

FIGURE 15.5   (See color insert.) Marketable onion bulb yield (kg/ha) as influenced by drip fertigation: 2011–12 and 2012–13.

Drip irrigation at 80% CPE resulted in a higher **pooled yield of marketable onion** (3833 kg/ha) followed by 60% CPE; whereas the lower pooled yield of marketable onion was recorded at 40% CPE. Among different fertilizer levels, fertilizer application at 100% RDF as WSF significantly produced a maximum pooled yield of marketable onion (4418 kg/ha) followed by 100% RDF (75% CF + 25%WSF) and 100% RDF as CF. Drip fertigation at 80% CPE with 100% RDF as WSF significantly recorded maximum pooled yield of marketable onion (4946 kg/ha) compared to other treatment combinations.

15.4.3   CASTOR EQUIVALENT YIELD

The crop equivalent yield of onion intercrop was converted into equivalent yield of castor (kg/ha) for both years. Castor equivalent yield (CEY) was significantly influenced by irrigation regimes and fertigation levels. Drip irrigation at 80% CPE of (castor + onion intercrop) recorded significantly higher castor equivalent yield (4866 kg/ha during 2011–12 and 4565 kg/ha during 2012–13, respectively) compared to 60% CPE (4176 and 3942 kg/ ha) during both years. The yield increase under 80% CPE irrigation regime was 16.5 and 15.8% than that at 60% CPE in both years. Drip irrigation at 40% CPE gave lower castor equivalent yield of 3986 kg/ha during 2011–12 and 3420 kg/ha and during 2012–13, respectively. Surface irrigation with soil application of 100% RDF as CF gave lower CEY (4744 and 4432 kg/ha) in both years than drip fertigation at 80% CPE with 100% RDF as conventional fertilizer. Among fertilizer levels, fertilizer at 100% RDF through WSF registered higher CEY of 5582 kg/ha and 5003 kg/ha during 2011–12 and 2012–13, respectively followed by 100% RDF (75% CF + 25% WSF) and 100% RDF through CF during both years. The percentage increase of CEY at 100% RDF with WSF was 10.6 and 11.3% during 2011–12 and2012–13, respectively than the fertilizer at 100% RDF as CF. The lower CEY of castor was recorded under drip fertigation at 40% CPE with 75% RDF through CF (3938 and 3586 kg/ha) during 2011–12, and 2012–13, respectively. The percentage yield increases in CEY of 18.1 and 12.9% was recorded under drip fertigation at 100% RDF with WSF compared to surface irrigation with soil application of 100% RDF as CF (the control) during 2011–12 and 2012–13, respectively. The interaction effects among irrigation regimes and fertigation levels were found significant. Drip fertigation at 80% CPE with 100% RDF through WSF application recorded significantly higher castor equivalent yield (6416 and 5759 kg/ha during 2011–12 and 2012–13, respectively) compared to other treatment combinations.

For pooled data, the 80% CPE resulted in better CEY of 4715 kg/ha than that at 60% CPE (4059 kg/ha) with yield increase of 16.2%. Lower CEY was recorded at 40% CPE (3703 kg/ha). Drip fertigation at 100% RDF as WSF gave 5292 kg/ha of CEY followed by 100% RDF (75% CF

+ 25% WSF). The lower CEY (3762 kg/ha) was recorded at 75% RDF as CF. The percentage increase in CEY was 10.9% in CEY at 100% RDF with WSF compared to 100% RDF as CF.

15.4.4 AGRONOMIC EFFICIENCY (KG (KG OF NUTRIENT)$^{-1}$)

The results showed that irrigation regimes and fertilizer levels had significant variation in agronomic efficiency (AE) of (castor + onion) intercrop. Among the irrigation regimes, drip irrigation at 80% CPE significantly gave higher AE (61.7 and 58.0 kg (kg of nutrient)$^{-1}$ during 2011–12 and 2012–13, respectively) compared to 60% CPE. At 40% CPE, AE was comparatively higher than surface irrigation during 2011–12 (51.3 kg (kg of nutrient)$^{-1}$) and lesser during 2012–13 (44.8 kg^{-1} of nutrient^{-1}), respectively. Agronomic efficiency was significantly higher at 75% RDF as WSF (58.6 and 55.0 kg (kg of nutrient) $^{-1}$ during 2011–12 and 2012–13, respectively) followed by 100% RDF as WSF during 2011–12; whereas it was followed by 75% RDF (75%CF + 25%WSF) during 2012–13, respectively. The AE showed significant interaction effect among irrigation regimes and fertilizer levels in both years. Application of 100% RDF as WSF with 40% CPE showed higher AE (66.1 kg (kg of nutrient) $^{-1}$) in 2011–12; whereas for 2012–13, higher AE was recorded at 40% CPE with 75% RDF as WSF (61.2 kg kg^{-1} of nutrient^{-1}) of (castor + onion) intercrop.

15.4.5 CROP WATER REQUIREMENT

In drip irrigation method, irrigation at 80% CPE consumed total amount of 388 and 456 mm of water during 2011–12 and 2012–13, respectively. Similarly, the treatment with 60% CPE irrigation water consumed total amount of 326 and 384 mm during both years. Drip irrigation at 40% CPE used lesser quantity of water (264 and 311 mm during 2011–12 and 2012–13, respectively). For the surface irrigation method, total often irrigations were given including sowing time and life irrigation. Total of 560 mm of water was used for the year 2011–12 with effective rainfall of 60 mm. Similarly, for 2012–13, the number of irrigations was nine consuming 539 mm of water with effective rainfall of 89 mm.

15.4.6   WATER USE EFFICIENCY (KG/HA-MM⁻¹)

For all irrigation regimes and fertilizer levels, significant variation was observed in water use efficiency (WUE) as shown in Tables 15.3 and 15.4.

TABLE 15.3   Water Applied, Effective Rainfall, Total Water Use and Water Use Efficiency of Castor + Onion Intercrop

Treatments	CEY (kg/ha)	Water applied (mm)	Effective rainfall (mm)	Total water use (mm)	WUE (kg/ha mm⁻¹)
			Rabi – 2011–12		
$I_1 F_1$	3634	224	40	264	13.8
$I_1 F_2$	4034	224	40	264	15.3
$I_1 F_3$	3824	224	40	264	14.5
$I_1 F_4$	4451	224	40	264	16.9
$I_1 F_5$	5082	224	40	264	19.3
$I_1 F_6$	4894	224	40	264	18.6
$I_2 F_1$	3805	286	40	326	11.7
$I_2 F_2$	4035	286	40	326	12.4
$I_2 F_3$	3940	286	40	326	12.1
$I_2 F_4$	4922	286	40	326	15.1
$I_2 F_5$	5247	286	40	326	16.1
$I_2 F_6$	5125	286	40	326	15.7
$I_3 F_1$	4376	348	40	388	11.3
$I_3 F_2$	4694	348	40	388	12.1
$I_3 F_3$	4619	348	40	388	11.9
$I_3 F_4$	5775	348	40	388	14.9
$I_3 F_5$	6416	348	40	388	16.5
$I_3 F_6$	6166	348	40	388	15.9
Surface irrigation	4744	500	60	560	8.5
			Rabi – 2012–13		
$I_1 F_1$	3141	245	67	311	10.1
$I_1 F_2$	3464	245	67	311	11.1
$I_1 F_3$	3256	245	67	311	10.5
$I_1 F_4$	3818	245	67	311	12.3
$I_1 F_5$	4231	245	67	311	13.6

TABLE 15.3   *(Continued)*

Treatments	CEY (kg/ha)	Water applied (mm)	Effective rainfall (mm)	Total water use (mm)	WUE (kg/ha mm^{-1})
$I_1 F_6$	4044	245	67	311	13.0
$I_2 F_1$	3504	317	67	384	9.1
$I_2 F_2$	4002	317	67	384	10.4
$I_2 F_3$	3809	317	67	384	9.9
$I_2 F_4$	4451	317	67	384	11.6
$I_2 F_5$	5018	317	67	384	13.1
$I_2 F_6$	4814	317	67	384	12.5
$I_3 F_1$	4112	390	67	456	9.0
$I_3 F_2$	4462	390	67	456	9.8
$I_3 F_3$	4471	390	67	456	9.8
$I_3 F_4$	5216	390	67	456	11.4
$I_3 F_5$	5759	390	67	456	12.6
$I_3 F_6$	5415	390	67	456	11.9
Surface irrigation	4432	450	89	539	8.2

TABLE 15.4   Effect of Drip Irrigation and Fertigation on Water Use Efficiency (kg/ha-mm^{-1}) of Castor + Onion Intercrop

Treatments	Rabi, 2011–12				Rabi, 2012–13			
	I_1	I_2	I_3	Mean	I_1	I_2	I_3	Mean
F_1 –75% RDF as CF	13.8	11.7	11.3	**12.2**	10.1	9.1	9.0	**9.4**
F_2 –75% RDF as WSF	15.3	12.4	12.1	**13.3**	11.1	10.4	9.8	**10.4**
F_3 –75% RDF (75% CF + 25% WSF)	14.5	12.1	11.9	**12.8**	10.5	9.9	9.8	**10.1**
F_4 –100% RDF as CF	16.9	15.1	14.9	**15.6**	12.3	11.6	11.4	**11.8**
F_5 –100% RDF as WSF	19.3	16.1	16.5	**17.3**	13.6	13.1	12.6	**13.1**
F_6 –100% RDF (75% CF + 25% WSF)	18.6	15.7	15.9	**16.7**	13.0	12.5	11.9	**12.5**
Mean	**15.1**	**12.8**	**12.6**		**11.0**	**10.3**	**10.0**	
Surface irrigation: 100% RDF as CF		8.5				8.2		

TABLE 15.4   *(Continued)*

	S Ed	CD (P= 0.05)	S Ed	CD (P= 0.05)
I	0.14	0.38	0.22	0.62
F	0.14	0.29	0.16	0.32
I x F	0.27	0.59	0.33	NS
F x I	0.25	0.51	0.27	NS

Among irrigation regimes, drip irrigation at 40% CPE resulted in significantly higher WUE (15.1 and 11.0kg/ha-mm^{-1} during 2011–12 and 2012–13, respectively), followed by 60% CPE and surface irrigation during both years. The WUE of surface irrigation was comparatively lower (8.5 and 8.2 kg/ha-mm^{-1}) than that at 80% CPE during both years. The WUE was decreased with increased level of irrigation. Significantly more WUE was recorded at 100% RDF as WSF application (17.3 and 13.1 kg/ha-mm^{-1} during 2011–12 and 2012–13, respectively), followed by 100% RDF (75% CF + 25% WSF) and 100% RDF as CF. Lower WUE of 12.2 and 9.4 kg/ha-mm^{-1} was recorded at 75% RDF as CF in both years. Interaction effect among irrigation regimes and fertilizer levels was significant on WUE during 2011–12. Significantly higher WUE was recorded under drip irrigation at 40% CPE with 100% RDF as WSF (19.3 kg/ha-mm^{-1}) during 2011–12 than other treatment combinations. There was no interaction effect during 2012–13.

15.4.7   EFFECTIVE UTILIZATION OF LAND AND WATER

Drip fertigation at 80% CPE with 100% RDF as WSF showed increased yield (35.3 and 30.0 during 2011–12 and 2012–13, respectively) followed by 80% CPE with 100% RDF (75% CF + 25% WSF) compared to surface irrigation with soil application of 100% RDF as CF during both years. Whereas at 60% CPE with 100% RDF as WSF showed increased yield of 10.6 and 13.2 compared surface irrigation with soil application of 100% RDF as CF during both years. There was a distinct difference in water use by surface irrigation method compared to drip irrigation method during both years. At 80% CPE, water saving was 30.8% during 2011–12, and 15.5% during 2012–13 compared to surface method of irrigation; whereas at 60% CPE, water saving was 41.8% and 28.9% during 2011–12 and

2012–13, respectively. With this saving of water, an additional area of 44.4 and 18.4% can be irrigated under drip irrigation at 80% CPE during 2011–12 and 2012–13, respectively.

15.4.8   SOIL MOISTURE DYNAMICS

The moisture movement varied in drip irrigation and surface method of irrigation, which in turn determines the wetting zone and effective root zone. Therefore, the soil moisture contents at varying radial and vertical distances were determined under drip– and surface irrigation methods.

Surface irrigation: In surface irrigation, soil samples were collected at 0 –15, 15–30 and 30 –45 cm depth at 48 hours interval after irrigation during post-rain free irrigation cycle and soil moisture depletion pattern was studied. The results showed that soil moisture depletion was faster during initial period of observations and was slowed down during later period and also just before irrigation. Unlike drip irrigation, the soil moisture content under surface irrigation method was steeply declined from 22.87% in top layer of 0–15 cm depth at 2 days after irrigation (DAI) to 12.14% on 14 DAI. Among different soil layers, the amount of moisture from the top layer was much higher than from deeper layer. The soil moisture content at surface layer in surface irrigation found lower compared to subsurface layers as the days after irrigation were increased and there was more depletion of moisture in 0–15 cm layer due to evaporation. When the depth of soil was increased, the moisture depletion rate was decreased. The soil moisture content under surface irrigation was comparable with drip irrigation treatments up to a period of 72 hours (18.46%), nearer to field capacity and thereafter there was a steady and steep decline in soil moisture content resulting in moisture stress.

Drip irrigation: Drip irrigation was scheduled once in three days (72 hours) with cumulative pan evaporation (CPE) for computed quantity of water as pretreatment schedule of 80, 60 and 40% CPE. The soil moisture content was estimated up to a depth of 30 cm across the lateral at 10 cm lateral and 20 cm along the lateral at 10 cm interval. The soil moisture content was estimated at 24 and 48 hours after drip irrigation. Moisture content estimated 24 hours after irrigation at various horizontal distances from the emitter location indicated that the moisture content was decreased in deeper soil layers as the distance from emitters was increased. The soil moisture content after 48 hours of drip irrigation up to 15 cm distance from dripper either along or across the lateral in all irrigation regimes was almost consistent and nearer to the field capacity. Under drip irrigation, the soil moisture content observed at 80% CPE was always maintained above 85% available soil moisture even at 30 cm distance across the lateral at 15–30 cm depth (18.28% for 24 hours after irrigation and 16.72% for 48 hours of irrigation). When application rate was decreased in drip irrigation, the moisture content was decreased distinctly with increased distance from dripper. Hence, it is evident that drip irrigation at reduced levels (*viz.*, 60 and 40% CPE) has resulted in lower soil moisture content near to plant compared to drip irrigation at 80% CPE.

15.4.9 NUTRIENT DYNAMICS UNDER DRIP FERTIGATION

The available N was medium at the end of fertigation cycle. The available nitrogen was increased steadily with increased distance from the dripper across and along to laterals up to a distance of 15 cm. The data revealed that at the end of fertigation at 100% RDF as WSF showed higher available N than at 100% RDF as CF. Among the distances, the peak available nitrogen was recorded at a distance of 15 cm from the dripper across the lateral. The nitrogen availability steadily was decreased with increased depth up to 30 cm distance. Across the laterals, the peak available soil nitrogen (243 and 237 kg/ha) was found at 100% RDF as WSF at the end of fertigation. It was present at depth of 15–30 cm at a distance of 15 and 30 cm from the dripper. Along lateral distance, the nitrogen concentration was steadily increased up to 20 cm distance from the dripper with 30 cm depth (239 and 248 kg/ha) at the end of fertigation for 100% RDF as WSF. The nitrogen content was decreased with decrease in RDF levels. The lower nitrogen concentration was recorded at fertigation with 75% RDF as conventional fertilizer.

Phosphorus dynamics: The distribution of available phosphorus in across and along lateral in the soil was influenced among fertigation levels. The higher available phosphorus in soil was confined to 0–15 cm of soil layer under all fertigation levels. The available P was decreased with increase in distance and depth. Higher P availability was recorded under higher RDF (100% RDF as WSF) and was decreased with decreased level of fertilizer dose. Application of 75% RDF as conventional fertilizer recorded lower available phosphorus content. The peak availability of phosphorus recorded just below the dripper (20.71 and 20.12 kg/ha, respectively at 0 and 15 cm distance from dripper) at 100% RDF as WSF at the end of fertigation.

Potassium dynamics: Potassium content was recorded at the end of last fertigation. At the end of fertigation cycle, K availability was lower. Available K content in the soil varied with layers and distance from the dripper point. Across the lateral, there was the higher available K at emitter point and was decreased as the distance increases with lower level to 30 cm depth. The peak availability of K availability was recorded just below the dripper (389 and 379 kg/ha at 0 to 15 cm depth just below the dripper) at 100% RDF as WSF and was decreased along and across the dripper.

15.4.10   ROOT STUDIES: CASTOR

Root weight: Root weight of castor was measured at the harvesting stage of crop. The root weight differed significantly among irrigation regimes and fertilizer levels in castor crop. Drip irrigation at 80% CPE measured the maximum root weight of 47.6 and 44.4 g followed by 60% CPE during 2011–12 and 2012–13, whereas the lower root weight was measured at 40% CPE, respectively. Among fertilizer levels, application at 100% RDF as WSF gave increased root weight of 56.2 and 52.2 g during 2011–12 and 2012–13, respectively; followed by 100% RDF (75% CF + 25% WSF) and 100 RDF as CF, whereas at 100% RDF (75% CF + 25% WSF) 100% RDF through CF was at par in root weight during both years. Whereas lower root weight (32.9 and 30.3 g) was observed at 75% RDF as CF in both years. Surface irrigation with soil application of 100% RDF as CF gave minimum root weight (46.3 and 41.7 g) compared to drip fertigation at 80% CPE with 100% RDF as CF in two years of study. Significant interaction between drip irrigation and fertilizer levels was examined in root weight in both years. Drip fertigation at 80% CPE with 100 RDF as WSF (72.7 and 67.5 g) significantly gave higher root weight in both years than other treatment combinations.

Root volume: Castor: Irrigation regimes and fertilizer levels significantly influenced the castor root volume during 2011–12 and 2012–13, respectively. Maximum root volume was recorded at 80% CPE (119.8 and 118 cm^3) compared to 60% and 40% CPE in both years. Among fertilizer levels, fertilizer application at 100% RDF as WSF showed maximum root volume (112.9 and 106.3 cm^3) followed by 100% RDF (75%CF + 25%WSF) and 100% RDF as CF in both years, respectively.

Minimum root volume was recorded at 75% RDF as CF application (84.8 and 80.0 cm^3) in two years of study. Surface irrigation with soil application at 100% RDF as CF gave root volume of 97.6 and 85.9 cm^3, respectively. Significant interaction between irrigation and fertilizer levels was observed in root volume during both years. Drip fertigation at 80% CPE with 100% RDF as WSF gave significantly maximum root volume (149.8 and 142.6 cm^3) in both years compared to other treatments.

15.4.11   ROOT LENGTH OF ONION INTERCROP

Root length of onion intercrop was found significant under irrigation and fertilizer levels during 2011–12 and 2012–13, respectively. Maximum root length of onion intercrop was recorded under drip irrigation at 40% CPE (6.6 and 5.6 cm). Minimum root length of 4.8 and 4.4 cm noticed at 80% CPE in both years. Fertilizer levels influenced the root length of onion during both years. Among fertilizer levels, application of 75% RDF as CF registered higher root length (6.1 and 5.4 cm) compared to other treatments in both years. Lower root length was recorded at 100% RDF as WSF in both the year of study. There was no interaction effect at drip fertigation in root length of onion during both years.

15.4.12   LAND EQUIVALENT RATIO (LER)

Land equivalent ratio indicates the relative land area under sole crop that is required to produce the yield achieved in intercropping system under same levels of management. "Castor + onion" intercropping under different irrigation regimes and fertilizer levels significantly influenced the land equivalent ratio (LER) both during 2011–12 and 2012–13 (Table 15.5).

Intercropping of "castor + onion" under drip irrigation at 80% CPE realized 22 and 19%yield advantage and recorded higher LER values of 1.22 and 1.19 during 2011–12 and 2012–13, respectively compared to sole crop of castor and onion. Among fertilizer levels, application of 100% RDF as WSF gave maximum LER value of 1.35 and 1.29 during 2011–12 and 2012–13, respectively compared to other fertilizer treatments during

TABLE 15.5　Effect of Drip Irrigation and Fertigation on Land Equivalent Ratio (LER) of Castor + Onion Intercrop

Treatments	Rabi, 2011–12				Rabi, 2012–13			
	I_1	I_2	I_3	Mean	I_1	I_2	I_3	Mean
F_1 –75% RDF as CF	0.88	0.92	1.09	**0.96**	0.71	0.86	1.07	**0.88**
F_2 –75% RDF as WSF	0.95	0.99	1.19	**1.04**	0.79	1.02	1.16	**0.99**
F_3 –75% RDF (75% CF + 25% WSF)	0.93	0.96	1.17	**1.02**	0.75	0.96	1.17	**0.96**
F_4 –100% RDF as CF	1.05	1.16	1.43	**1.21**	0.89	1.09	1.35	**1.11**
F_5 –100% RDF as WSF	1.21	1.26	1.59	**1.35**	1.07	1.29	1.52	**1.29**
F_6 –100% RDF (75% CF + 25% WSF)	1.16	1.21	1.54	**1.31**	0.99	1.22	1.41	**1.21**
Mean	**0.95**	**1.01**	**1.22**		**0.79**	**0.98**	**1.19**	
Surface irrigation: 100% RDF as CF		1.21				1.16		

	S Ed	CD (P= 0.05)	S Ed	CD (P= 0.05)
I	0.02	0.06	0.02	0.01
F	0.01	0.03	0.01	0.03
I x F	0.03	0.08	0.03	0.06
F x I	0.02	0.05	0.02	0.04

both years. Lower LER value of 0.96 and 0.88 was recorded at 75% RDF as CF during 2011–12 and 2012–13, respectively.

15.4.13　COST ECONOMICS OF CASTOR + ONION INTERCROPPING SYSTEMS

The data on economics of drip fertigation for castor + onion intercropping are presented in Appendices I to II. The lifespan of drip system varies from 8 to 10 years depending upon quality and maintenance of drip system. Hence a normal lifespan of 10 years was considered for computation. Though the initial capital investment was high for drip irrigation system, yet the cumulative benefit would be greater, considering the longer life of system. The fixed cost towards drip system installation was Rs. 74,583 per ha (1243 US$/ha) and the annualized cost interest was only Rs. 14,425 ha^{-1} (241 US$/ha) including repair and maintenance costs.

Economic analysis of present investigation revealed that the cost of cultivation for drip irrigation was higher than surface irrigation irrespective of fertilizer levels. In general, the cost of cultivation ranged from Rs. 67,508 per ha (1125 US$/ha) at 100% RDF as WSF and Rs. 55,430 ha^{-1} at 75% RDF as CF under drip fertigation system compared to Rs. 55,027 per ha (917 US$/ha) for surface irrigation with soil application of 100% RDF as conventional fertilizer.

Drip irrigation at 80% CPE recorded higher net income and it was followed by drip irrigation at 60% CPE. Surface irrigation with soil application at 100% RDF as CF was comparatively higher than 40% CPE with 75% RDF as CF. With respect to 100% RDF as WSF, drip irrigation scheduled at 80% CPE resulted in higher income of Rs. 1,42,632 per ha (2377 US$/ha) during 2011–12 and Rs.1,19,642 per ha (1994 US$/ha) in 2012–13, respectively followed by drip fertigation at 100% RDF (75%CR + 25%WSF) compared to surface irrigation with soil application of 100% RDF as CF during both years. Surface irrigation with soil application of 100% RDF as CF resulted in higher income of Rs. 1,11,018 per ha (1850 US$/ha) during 2011–12 and Rs. 1,00,108 per ha during 2012–13, respectively.

Lower income was attained at 40% CPE with 75% RDF as CF and it was Rs. 57, 350 per ha (956 US$/ha) during 2011–12 and Rs. 40,080 per ha (668 US$/ha) during 2012–13, respectively.

Among the various drip fertigation treatments, drip fertigation at 80% CPE with 100% RDF (75%CF + 25%WSF) gave maximum net return of Rs. 1,63,921 per ha (2732 US$/ha) during 2011–12 followed by 80% CPE with 100% RDF as WSF Rs. 1,61,449 per ha (2690 US$/ha); whereas in 2012–13 drip fertigation at 80%with 100% RDF as WSF resulted in maximum net return (Rs. 1,32,840 per ha (2214 US$/ha)) followed by 80% CPE with 100% RDF (75%CF + 25%WSF) (Rs.1,31,223 per ha (2020 US$/ha)) compared to other treatments. Lower net return of Rs. 50,747 per ha (846 US$/ha) in 2011–12 and Rs. 29,659 per ha (494 US$/ha) in 2012–13 was realized at 40% CPE with 75% RDF as CF, respectively.

The benefit-cost ratio (BCR) for various treatments revealed that surface irrigation recorded higher benefit-cost ratio of 3.02 followed by 80% CPE with 100% RDF (75%CF + 25%WSF) with benefit-cost ratio of 2.71 and drip irrigation at 80% CPE with 100% RDF as CF (2.63) during 2011–12, whereas during 2012–13, surface irrigation registered higher BCR of

2.82 followed by drip irrigation at 80% CPE with 100% RDF (75%CF + 25%WSF) value of 2.33. The lower benefit-cost ratio value of 1.45 and 1.29 was recorded at 40% CPE with 75% RDF as WSF in both years.

The net present value and discounted benefit-cost ratios were worked out for all drip fertigation treatments to determine viable economic combination. Drip fertigation at 80% CPE with 100% RDF as WSF resulted in higher net present value of Rs. 97,019 per ha (1616 US$/ha) and discounted benefit-cost ratio of 2.30 followed by drip fertigation at 80% CPE with 100% RDF (75% CF + 25% WSF) with net present value of Rs. 85,014 per ha (1417 US$/ha) and discounted benefit-cost ratio of 2.14. Whereas, drip fertigation with 100% RDF as CF resulted in net present value of Rs. 34,468 per ha (574 US$/ha) and discounted benefit-cost ratio of 1.46.

15.5   DISCUSSION

15.5.1   CASTOR CROP

15.5.1.1   EFFECT OF WEATHER PARAMETERS ON THE PERFORMANCE OF CROPS

The prevalent weather conditions were conducive for the normal growth and development of castor and onion, which had enabled a favorable growth without any adverse effect on the incidence of pest and diseases. The major part of rainfall was received during early growth stage of crop during both the seasons and this had remarkable influence on the availability of applied nutrients and consequently enhanced their uptake thus promoting the growth and yield attributing characters. Among two years of study, "castor + onion" intercrop raised during 2011–12 received solar radiation of 436.7 Cal cm^{-2} day^{-1} and wind velocity of 4.8 km h^{-1} and mean pan evaporation of 4.1 mm/day and were optimum throughout the growing period, which reflected on higher growth parameters, yield attributes and subsequently the crop yield compared to the crop raised in 2012–13 with mean solar radiation of 443 Cal cm^{-2} day^{-1} and 5.4 km h^{-1} with higher pan evaporation (4.5 mm day^{-1}). These results are in accordance with the findings of Mingochi (1998), who stated that the high wind velocity increases evapotranspiration and crop water requirement with resultant yield reduction. Annandale et al. (2004) also reported that very high solar radiation may cause detrimental effect on crop and may result in lower yield.

15.5.1.2   EFFECTS OF DRIP FERTIGATION ON GROWTH PARAMETERS OF CASTOR

Significant improvement on growth parameters (such as days to 50% flowering, plant height, dry matter production (DMP) and leaf area index (LAI)) was observed due to irrigation regimes and fertilizer levels during 2011–12 and 2012–13. All growth components were positively influenced due to varying irrigation regimes and nutrient levels at all growth stages of crop. The treatment combination of scheduling irrigation through drip at 80% CPE and fertigation with 100% RDF as WSF took more number of days to 50% flowering during both years, due to availability of optimum moisture conditions and more available nutrients, which prolonged the vegetative growth of castor, which in turn favored better yield. Similar results were reported by Sree and Reddy (2003).

Drip irrigation regime at 80% CPE with 100% RDF as WSF as fertigation at 90 DAS recorded higher plant height during 2011–12 and 2012–13 (Figure 15.6), due to adequate available soil moisture in root zone, which resulted in better root development and positive absorption of nutrients, use of solar radiation and natural resources more efficiently that consequently accelerated the photosynthesis rate by increase in protein molecules, amino acids and nucleotides of crops (Figures 15.6 and 15.7). Similar results were reported by Giridhar and Gajendragiri (1991).

Optimum temperature and solar radiation were most important factors for increasing dry matter production (DMP). Application of 100% RDF as WSF with drip irrigation at 80% CPE significantly recorded higher dry matter production at various growth stages during 2011–12 and 2012–13 (Figure 15.7) and this was mainly due to optimum moisture and timely nutrient application, which could have enhanced the assimilatory efficiency resulting in increased number of nodes per plant, better branching, and LAI that contributed to higher biomass production.

Scheduling irrigation through drip along with fertigation could have promoted the activity of photosynthesis and simultaneous accumulation of dry matter. Similar findings of increased DMP under drip fertigation were reported by Randhawa and Venkateswarulu (1980).

Leaf area index (LAI) is one of the principle factors influencing canopy net photosynthesis (Hansen, 1972). Higher LAI is positively correlated with yield and efficient utilization of resources. At various growth stages, drip irrigation at 80% CPE along with fertigation at 100% RDF as WSF

gave significantly superior LAI compared to conventional surface irrigation during both years. Increase in LAI due to maximum growth expres-
sion, more number of leaves increased size of leaves, and favorable
leaf orientation might be due to available optimum moisture along with
required nutrients that created favorable conditions to put forth optimum
growth and better development of yield attributing characters of castor.

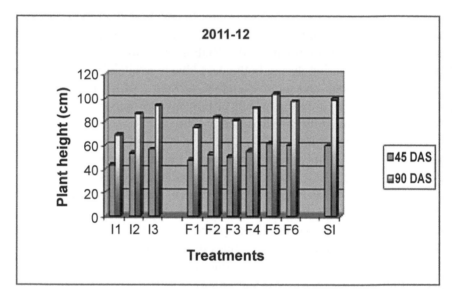

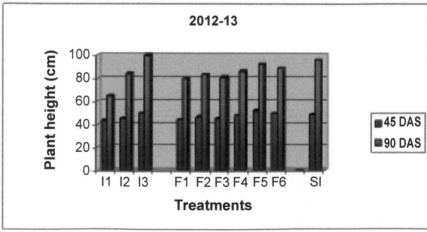

FIGURE 15.6 (See color insert.) Effects of drip fertigation on plant height (cm) of
castor: 2011–12 and 2012–13.

The crop growth rate (CGR) was increased with increasing levels of irrigation and fertilizer doses in castor + onion intercrop. Combination of drip irrigation at 80% CPE and fertigation with 100% RDF as WSF gave higher CGR during both years. It might be due to utilization of water and nutrient resources effectively in the mechanism of cell division, and elongation for every incremental increase in crop growth.

Krieg and Sung (1986) reported that water stress caused a reduction in the whole plant leaf area by decreasing the initiation of new leaves, with no significant changes in leaf size. Both the main and side branches developed significantly less leaves; however, the effect was less severe on the main stem leaves. Pettigrew (2004) reported that water deficit stress resulted in decreased leaf size, but this decrease was accompanied by an increase in the specific leaf weight (SLW), a phenomenon observed by Wilson et al. (1987). Specific leaf weight showed positive, significant relationships with yield. This shows that selection based on specific leaf weight have resulted in increased yield. The increased yield of trait under stress conditions is caused by increase in number of mesophyll cells per unit area (Ober et al., 2005). Drip irrigation at 40% CPE with 75% RDF as CF combination registered maximum specific leaf weight. The results obtained in the current study could be related to the rapid translocation of assimilates to the sink under the conditions of leaf death rate, which occurs in stress condition.

Relative water content (RWC) represents the ability of the crop to retain tissue water status under water stress and the castor hybrid retaining more leaf tissue are expected to perform better. Castor can withstand drought and maintain leaf water level under adverse climatic condition. Mild water stress (80% CPE) with 100% RDF as WSF gave significantly higher RWC by maintaining leaf relative water content percentage at higher level at vegetative stage, primary spike stage and tertiary spike stages during both years. With incremental of fertilizer application, the RWC resulted in varying crop growth stages. Under stress (40% CPE), RWC was decreased compared to mild water stress. Decrease in RWC due to moisture stress in cotton has also been reported (Janagoudar et al., 1983).

15.5.1.3   EFFECT OF DRIP FERTIGATION ON YIELD ATTRIBUTES OF CASTOR

Yield in castor is a highly variable factor and varies with genotype, cultural, and management practices, especially water and fertilizer application. The

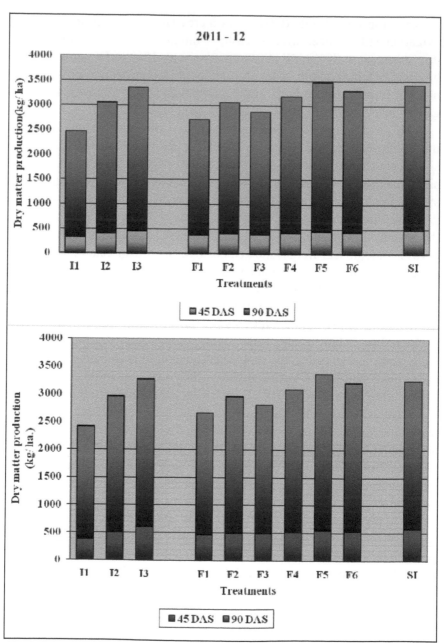

FIGURE 15.7   (See color insert.) Effects of drip fertigation on dry matter production (DMP, kg/ha) of castor: 2011–12 and 2012–13.

yield was favorably influenced by irrigation regimes and different levels of nutrients, the effect had been expressed in terms of yield components *viz.*, length of primary spike, number of capsules per plant, total number of spikes plant^{-1} and 100 seed weight (Figure 15.8).

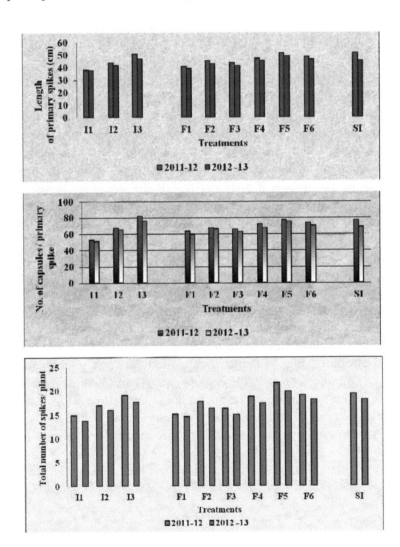

FIGURE 15.8 **(See color insert.)** Effects of drip fertigation on yield attributes of castor: 2011–12 and 2012–13.

In general, crops respond well to water regimes and fertilizer levels and castor is not an exception. At 80% CPE, the yield attributes (*viz.*, length of primary spike, number of capsule plant^{-1}, total number of spikes plant^{-1} and 100 seed weight) registered significantly higher values. Optimum dose of fertilizer application is very important for getting higher yields of castor and has been emphasized by Mathukia and Modhwadia (1995).

Application of 100% RDF as WSF recorded significantly higher primary spike length, total number of capsules per primary spike, and 100 seed weight at 90 DAS. Total number of spikes per plant was significantly higher under different stages during both years.

Higher values of yield attributes were recorded under best treatment combination of 80% CPE with 100% RDF as WSF, due to higher rate of photosynthesis, which resulted in better plant development, seed filling, and consequently higher seed yield. Nitrogen application with irrigation enhanced the production of male and female flowers. Thus there were an increased number of capsules per plant, which ultimately increased the yield.

Increased values of yield attributes might be also due to adequate available soil moisture, which in turn increased the cell turgor leading to effective physiological processes like cell division, cell elongation, better root development for more nutrient absorption and profuse vegetative growth, which resulted in higher setting. The rate of appearance of floral primordial is increased under adequate moisture status and all these factors lead to more seed settings. Similar results were reported by Rao and Venkateshwarlu (1988). Higher dose of fertilizer application showed improvement in yield attributes, due to better coordination between sources and sink activity, as nitrogen is one of the components in carbohydrate metabolism. Increase in test weight might be due to bold and well-developed seeds as a result of better availability of water and nutrient at higher levels. These findings are in agreement with the results of Rizaddin Ahamed et al. (2001).

15.5.1.4   EFFECT OF DRIP FERTIGATION ON SEED YIELD OF CASTOR

Drip irrigation regime at 80% CPE with 100% RDF as WSF significantly produced maximum seed yield of castor (Figure 15.4), due to availability of sufficient moisture and efficient utilization of nutrients. According to Pratap Kumar Reddy et al. (2006), irrigation has to be scheduled at 55 mm to 80 mm CPE for winter castor. Naghabhushanam and Raghavaiah (2005) also concluded that castor has to be irrigated at 0.8 IW/CPE ratio in wet regime. Sesha-Saila and Baskar Reddy (2005) recommended that irrigation scheduling at 80 mm CPE for summer castor under rice fallow gave significantly higher seed yield of castor.

Drip fertigation at 100% RDF as WSF resulted in significant improvement in seed yield over other drip fertigation with onion intercropping, due to better utilization of nutrient owing to good development of root system of castor crop. The results confirm the finding of Patel et al., (2006). It also produces better vegetative structure for nutrient absorption, strong sink strength through development of reproductive structures and production of assimilates to fill economic important sink. The results corroborate the findings of Rana et al. (2006). Intercropping of onion with castor causes vigorous growth without stiff competition because of efficient utilization of resources (water and nutrient) during both years. There was no significant yield reduction in castor due to intercropping with black gram or green gram or soybean according to Srilatha et al. (2002).

15.5.1.5   EFFECT OF DRIP FERTIGATION OF NUTRIENT UPTAKE: CASTOR

Total nitrogen (N), phosphorus (P_2O_5) and potassium (K_2O) uptake was significantly influenced due to varying irrigation and fertilizer levels at all growth stages.

Irrigation regime at 80% CPE and fertigation at 100% RDF as WSF registered higher uptake of N compared to other treatments. The improvement in N-uptake of nutrients for increasing the yield might be due to more vegetative structure for nutrient absorption, strong sink strength through development of reproductive structure and production of assimilates to fill economic important sink. The results corroborate the findings of Rana et al. (2006).

Phosphorus is a plant nutrient involved in a wide range of plant processes from cell division to the development of a good root system and ensuring timely and uniform ripening. It also performs number of functions related to growth, development, photosynthesis, and utilization of carbohydrates (Tandon, 1991). The increase in castor equivalent yield may be attributed by enrichment of soil with P, resulting in more P-uptake. In the present study, all the processes were favorably improved with application of 30 kg P_2O_5 ha^{-1}. The present findings corroborate the results of Raghavaiah (1999). The significant effect of P was observed in castor equivalent yield in castor + coriander intercropping system by Aglave et al. (2010).

Potassium applied (30 kg/ha) under drip fertigation at 80% CPE with 100% RDF as WSF significantly increased castor yield compared to other treatment combinations. This might be due to increased uptake of potassium and this can be ascribed to the influence of applied K_2O on the availability of K_2O in the soil and its extraction by plants as well as concomitant increase in seed yield. The present finding corroborates the results of Verma et al. (2010).

The higher N, P, and K removal by castor under higher levels of nutrient was probably reflected in dry matter production and seed yield. In addition to these, N fertilization had increased the cation exchange capacity of plant roots and thus made them more efficient in absorbing nutrient ions as suggested by Mathukia and Modhwadia (1995). Castor root has capacity to extract phosphorus from not easily soluble calcium phosphate. Similar views have been expressed by Paida (1976).

15.5.2 ONION INTERCROP

Understanding the wider row spacing of castor under drip irrigation situation, onion is mainly grown as intercrop to utilize the resources efficiently to complete the life cycle of castor main crop.

15.5.2.1 EFFECT OF DRIP FERTIGATION ON GROWTH AND PHYSIOLOGICAL PARAMETERS OF ONION INTERCROP

Higher plant height of onion intercrop was recorded under drip irrigation at 80% CPE with 100% RDF as WSF when compared to surface irrigation with soil application of 100% RDF as CF during both years. This might be due to higher irrigation regimes maintaining most of the root zone at adequate soil moisture content that did not fluctuate between wet and dry regimes. This result corroborates with the findings of Patil and Janawade (1999). Balasubramanyam (2003) also reported that onion plant height was increased with increased amount of irrigation water. Sufficient supply of nutrients showed stimulatory action in terms of cell elongation and thus resulting in increased plant height.

In the present study, number of leaf sheaths plant^{-1} was significantly higher under drip irrigation at 80% CPE and 100% RDF as WSF compared to 40% CPE. It might be due to better sink developed by continuous supply of nutrients under higher irrigation levels (Maya, 1996). Meenakshi and Vadivel (2003) also observed similar results of maximum number of leaf sheaths plant^{-1} with 100%water soluble fertilizer. It might be due to high level of N, P, and K during early stage, which would have increased the root activity. The transport of cytokinins from the root would have encouraged the growth and increased the number of leaf sheaths. These results are in agreement with Pandey et al. (1996).

The biological efficiency of any species could be reflected in the dry matter production. It is, therefore, appropriate to state that onion plant is intensifying dry matter accumulation after bulb differentiation. This observation is in accordance to Shadbolt and Holm (1956). Higher dry matter production was recorded under drip irrigation at 80% CPE than other drip irrigation regimes and surface irrigation. This might be due to increased plant height and more number of leaf sheaths as a result of maintenance of favorable soil moisture in the root zone. This is in line with the findings of Satyendrakumar et al. (2007), who found that biomass dry weight per plant was increased with the amount of water applied and varied between the treatments directly due to variation in vegetative growth parameters. Maximum DMP occurred at 100% RDF as WSF than other fertilizer levels. This might be due to the fact that nitrogen might be responsible for enhancing the photosynthetic ability while better availability and absorption of potassium could have helped in translocation of metabolites (especially sugars and carbohydrates) to the sink and thereby it increases the growth. This is in agreement with the earlier works of El-Sherif et al. (1993) in tomato.

Effects of water deficit on physiological processes are three dimensional. The first order processes affected by deficit are cell expansion, mesophyll resistance and stomatal resistance followed by second-order processes (leaf growth rate and rate of photosynthesis at leaf level and so on until yield is affected). The third dimension is related to timing of deficit in life cycle of the plant. As canopy approaches, closure, further cell expansion, leaf growth, and leaf area expansion become less important as determinants of yield that can be affected by water deficit (Hearn, 1994).

The CGR was increased with increased level of drip irrigation and fertilizer levels. Drip irrigation at 80% CPE with fertilizer at 100% RDF as WSF gave significantly higher CGR, due to continuously wetting moisture in the root zone thus influencing nutrient absorption, which enhanced the crop growth rate. This is in accordance with the findings of Shamima et al. (2003). Figure 15.9 indicates effects of drip fertigation on dry matter production on onion intercrop.

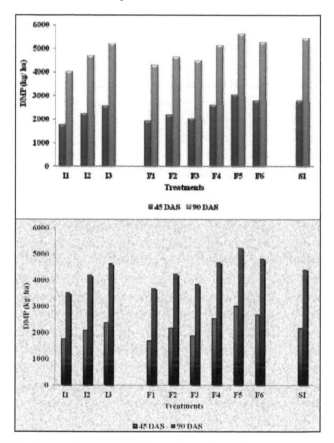

FIGURE 15.9 **(See color insert.)** Effects of drip fertigation on dry matter production (kg/ha) of onion intercrop, during 2011–12 and 2012–13.

15.5.2.2   EFFECT OF DRIP FERTIGATION ON YIELD ATTRIBUTES OF ONION INTERCROP

Yield is a complex trait influenced by morphological, physiological, and yield parameters. The ultimate goal of any management practice is to improve the yield level with minimum cost of production. In onion, the important yield contributing characters are equatorial and polar diameters, number of bulbs plant[-1], single bulb weight (g), and total bulb weight (g) plant[-1].

Higher yield attributes obtained at 80% CPE might be because of continuous and uninterrupted supply of soil moisture and nutrient to the crop (Figures 15.10a and 15.10b). The increase in bulb diameter due to drip irrigation has been observed by Ansary et al. (2006). Bybordi and Malakouti (2003) reported that onion diameter and weight were improved by increasing levels of Potassium application. The bulb weight was increased with increased level of fertilizer dose. Bulb weight was higher at 1.20 CPE as reported by Woldetsadik et al. (2003). Onion plant experienced water stress during the growing period: bulb development stages at 40% CPE, for which the bulb weight was significantly declined.

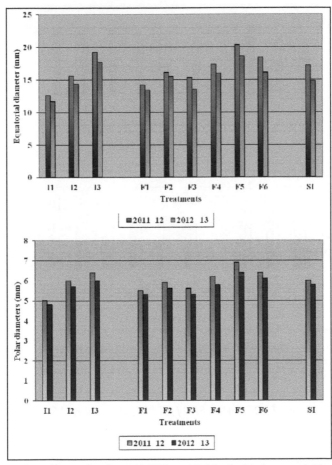

FIGURE 15.10a (See color insert.) Effects of drip fertigation on yield attributes of onion intercrop, during 2011–12.

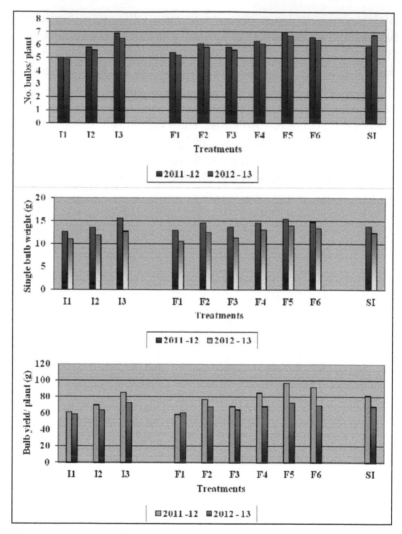

FIGURE 15.10b   (See color insert.) Effects of drip fertigation on yield attributes of onion intercrop, during 2012–13.

15.5.2.3   EFFECT OF DRIP FERTIGATION ON MARKETABLE ONION BULB YIELD

The aim of the onion intercropping with castor study is to get increased yield. Higher yield was recorded under drip irrigation at 80% CPE with fertilizer application at 100% RDF as WSF compared to other levels of drip fertigation and surface irrigation. Yield increase of marketable onion bulb under drip irrigation at 80% CPE during both the years was mainly due to consistent moisture availability that leads to quick growth, more dry matter production and better yield attributing characters.

Micro irrigation system recorded higher bulb onion yield than surface irrigation according to Corgan and Kedar (1990). Application of 100% RDF as WSF resulted in higher bulb yield of onion compared to other fertigation levels. This may be due to fertigation with higher fertilizer resulting in higher availability of all major nutrients in soil solution, which led to increased growth and leaf area, higher uptake and better translocation of assimilates from source to sink that in turn increased the bulb yield. Sankar et al. (2005) also observed higher crop yield of onion with higher dose of fertilization.

15.5.2.4 EFFECT OF DRIP FERTIGATION ON NUTRIENT UPTAKE BY ONION

Vegetable crops differ widely in their macronutrient requirements and in patterns of uptake during the growing season. In general, N, P, and K-uptake follow the same course as the rate of crop biomass accumulation. Drip fertigation not only acted as source of nutrients but also influenced their availability.

Drip irrigation at 80% CPE produced significantly higher nutrient uptake by onion intercrop, due to higher nutrient availability and greater absorption of nutrients by plant at optimum soil moisture. This is in conformity with the findings of Black (1969), who reported an increased nutrient uptake under higher frequency irrigation, due to increased plant growth. Bobade et al. (2002) reported that higher rates of N resulted in better translocation of assimilates from source to sink. The ready and continuously available moisture to plants in drip irrigation method might have also helped in increasing the total N-uptake as observed by Bafna et al. (1993). The increase in P-content with applied N may be due to the increased vegetative growth leading to higher P-uptake under favorable circumstances under increased level of drip irrigation. Heyman and Mosse (1972) also reported similar findings. The reason for higher concentration of K might be the consequence of higher demand of the expanding foliage and increased absorption to maintain the growth. Increased dose of fertilizer (100% RDF as WSF) was contributed for more K content in plant. These results are in conformity with the findings of Fontes et al. (2000).

15.5.3 EFFECT OF DRIP FERTIGATION ON CASTOR + ONION INTERCROP YIELD

The special configuration of "castor + onion" intercrop grown in water regimes and fertigation levels had significantly influenced castor equivalent yield (CEY) during both years. Drip irrigation under 80% CPE with application of 100% RDF with WSF significantly gave higher castor equivalent yield compared to other treatment combinations during both years. This might be due to optimum moisture nearer to field capacity which favored the nutrient uptake of castor and onion intercrop and thus simultaneously increasing the CEY.

Intercropping of castor with cluster bean and cucumber systems exploited resources in a better way to increase the system productivity. However, there was no significant reduction in castor yield due to intercropping with black gram or green gram as reported by Srilatha et al. (2002). Hence it did not exert stiff competition for the resources on castor crop during both years. Significantly similar castor grain equivalent was recorded in castor + coriander intercropping system. These results were in conformity with those by Gangasaran and Gajendra Giri (1983).

15.5.4 EFFECT OF DRIP FERTIGATION ON NUTRIENT USE EFFICIENCY AND AGRONOMIC EFFICIENCY OF CASTOR + ONION INTERCROP

The nutrient use efficiency was well pronounced by drip irrigation and fertigation levels in both years of experimentation. The agronomic efficiency (AE) was decreased with decreased level of drip irrigation. Drip irrigation at 80% CPE registered higher AE of nutrient. This might be due to uniform distribution of fertilizer with minimum leaching beyond the root zone. This confirms the findings of Bharambe et al. (1997), who mentioned the decrease in nutrient use efficiency with increased level of nutrients. Mohammad (2004) reported that AE was decreased with increased rates of fertigation.

15.5.5   EFFECT OF DRIP FERTIGATION ON WATER REQUIREMENT AND WATER USE EFFICIENCY OF CASTOR + ONION INTERCROP

The evapotranspiration and demand of water for metabolic activity of a crop constitute the consumptive use of water including the effective rainfall during the crop growing season. During both the years of study, consumptive use of water was higher under drip irrigation at 80% CPE compared to surface irrigation (Figure 15.11). This corroborates with the findings of Bekele and Tilahun (2007), who reported lower yield with greater water use efficiency under low levels of drip irrigation. Surface irrigation resulted in low water use efficiency. This might be due to decreased root conductance of water under deficit conditions inducing stomatal closures and less transpiration, which lead to poor water use efficiency. This is in line with findings of Drew and Lynch (1983).

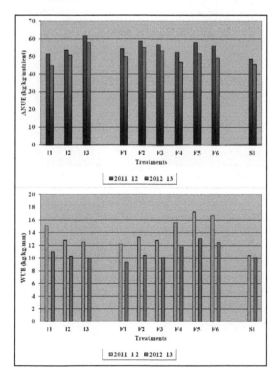

FIGURE 15.11   (See color insert.) Effects of drip fertigation on agronomic efficiency and water use efficiency of castor + onion intercrop: 2011–12 and 2012–13.

15.5.6   SOIL MOISTURE DISTRIBUTION

Drip irrigation once in three days had higher moisture content in the root zone depth (15–30 cm) in all regimes. Generally, the soil moisture content was decreased with increased distance from the laterals (Figures 15.12 and 15.13).

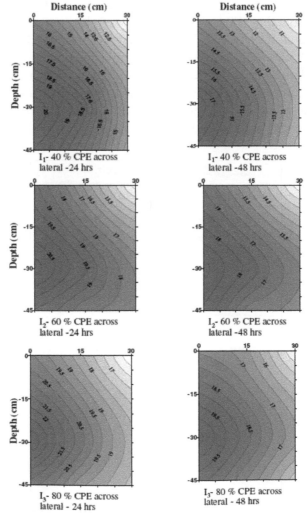

FIGURE 15.12   (See color insert.) Soil moisture (% weight basis) distributions across drip lateral for castor + onion intercrop.

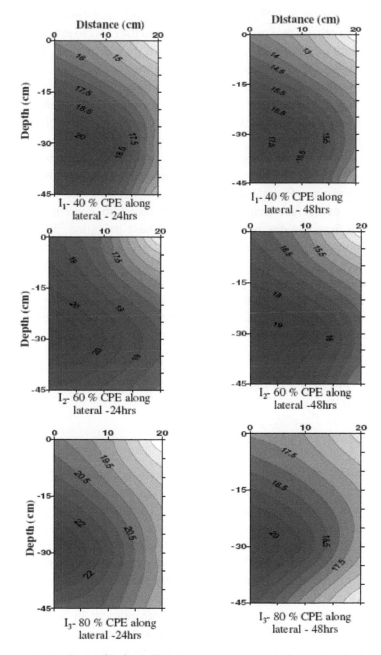

FIGURE 15.13   (See color insert.) Soil moisture (% weight basis) distributions along the drip lateral, for castor + onion intercrop.

Drip irrigation at 80% CPE recorded higher moisture content when compared to other water regimes. This enabled the roots to extract adequate soil moisture continuously during the entire growth period. This was in conformity with the findings of Satyendrakumar et al. (2007).

Soil moisture content just below the emitter (0 cm away from the dripper) was nearer to field capacity. The soil moisture content at 30 cm distance away from dripper was comparatively lower. Soil moisture distribution mainly depended on the rate of application and amount of water, initial moisture content of the soil (Khepar et al., 1983). The soil moisture decreased as the distance from emitter increased. This was also indicated by Sivanappan and Padmakumari (1980).

In case of surface irrigation, the soil moisture content was highly fluctuating from the day of irrigation until next irrigation. The irrigation interval between two successive surface irrigations was higher causing fluctuations in field capacity to stress conditions. This steep decline in soil moisture content made the plant roots hard enough to extract moisture. Similar results were reported by Senthilkumar (2000). These two extremes of moisture availability caused poor physiological activity of the crop, which reflected on poor growth and yield.

15.5.7   EFFECT OF DRIP FERTIGATION ON NUTRIENT DYNAMICS

15.5.7.1   NITROGEN DYNAMICS

The mobility of nutrient was well pronounced under drip fertigation system. In all drip irrigation levels, the nitrogen concentration in soil was increased from the emitter up to a certain distance and then was declined thereafter. The peak nitrogen concentration was recorded at 15–30 cm depth and 15 cm distance away from the dripper. Urea is relatively mobile and not strongly adsorbed by soil colloids. It tends to be more evenly distributed down the soil profile below the emitter and had moved laterally in the profile to 15 cm radius from the emitter (Haynes, 1990). Data from the present study showed that nitrogen content in soil profile has neither accumulated at the periphery of the wetting front nor was leached from the root zone, thus confirming findings of Chakraborty et al. (1999). The maximum concentration of nitrogen was noticed under higher level of drip fertigation (e.g. 80% CPE with 100% RDF as WSF). When the recommended dose of fertilizer was reduced, the availability of N was decreased in the root zone.

15.5.7.2   PHOSPHORUS DYNAMICS

In the present study, the movement of phosphate ion form from the emitter is very much dependent upon the adsorption capacity of the soil. Raushh-kolb et al. (1978) observed considerable vertical and horizontal movement of phosphate in clay loam soil. Unlike Nitrogen, the higher concentration of Phosphorus was seen at 0 –10 cm soil layer than at 10 –20 and 20–30 cm at the distance from the dripper. The restricted mobility of phosphorus might be due its strong retention by soil colloids and clay minerals. This result was in conformity with the findings of Harjindersingh et al. (2004). Phosphorus is less mobile in the soil and tends to accumulate under the dripper, with a little being leached downward or moved laterally (Alva and Syvertsen, 1991).

15.5.7.3   POTASSIUM DYNAMICS

Potassium is less mobile than nitrate but distribution in the wetted volume may be more uniform due to interaction with binding sites (Kafkafi et al., 1988). In the present investigation, the distribution of Potassium varied vertically and horizontally from emitting point and the movement of Potassium indicated a decreasing trend with respect to the soil depth, which is in accordance with the findings of Singh et al. (2000). The peak quantity of K was observed at 0 –15 cm depth under the emitter in the treatment receiving 100% RDF as WSF due to deposition of higher quantity in the upper layers. This falls in line with the findings of Singh et al. (2002). In this study, soil K content was significantly higher in the surface soil than in the subsoil and this might be due to majority of applied K was held in the surface soil and that downward movement of K was slower. Slow downward movement of applied K might be partially attributed to net upward flux of soil water in profile as a result of high evapotranspiration in summer. This is in line with the findings of Zeng et al. (2000).

15.5.8 EFFECT OF INTERCROPPING ON LAND EQUIVALENT RATIO

Intercropping of castor + onion significantly recorded higher land equivalent ratio (LER) under drip fertigation at 80% CPE with application of 100% RDF as CF during both years and a progressive increase in LER was observed with increasing levels of irrigation through drip and higher levels of nutrient application as WSF. This may be attributed to better utilization of moisture and nutrients, which might have helped for efficient utilization of natural resources by the component crops under intercropping system and enabled to yield better which ultimately resulted in higher land equivalent ratio. Reddy (1994) reported that pigeon pea, sunflower, and cowpea were found compatible with castor as evidenced by higher land equivalent ratio than unity, while sesame was found adversely affecting the crop with LER value less than unity. The increased yield of intercropping system might be the plausible reasons for such increase in LER under cassava + cowpea intercropping systems (Mohamed-Amanullah et al., 2006).

15.5.9 EFFECT OF DRIP FERTIGATION ON COST ECONOMICS OF CASTOR + ONION INTERCROPPING SYSTEM

Drip irrigation at 80% CPE with 100% RDF (75%CF + 25%WSF) resulted in higher gross net return and benefit-cost ratio (BCR) than other drip fertigation and surface irrigation with same level of fertilizer dose as soil application. Drip fertigation at 80% CPE with 100% RDF (75%CF + 25%WSF) was found to be economical because of lesser expenditure and equivalent yield as that of water-soluble fertilizer (WSF).

Higher net income was recorded at 80% CPE with 100% RDF (75%CF + 25%WSF) during 2011–12, whereas during 2012–13, 80% CPE with 100% RDF as WSF recorded higher net income compared to other treatments. This might be due to higher cost of fertilizer.

Drip irrigation at 80% CPE with 100% RDF as WSF resulted in higher net present value and discounted benefit-cost ratio

Although surface irrigation at 100% RDF as soil application gave higher BCR, yet the poor water and agronomic use efficiencies and low yield are drawbacks compared to drip irrigation practices. Also, higher net income per unit water consumption and net extra income over conventional method of irrigation for either one-hectare basis or equal water usage are more promising in drip irrigation. Basith and Mohammad (2010) reported that higher net income was recorded under drip irrigation over conventional irrigation methods. The intercropping emerges as statistically superior system with higher crop equivalent yields than sole cropping system. Incidentally these intercropping also fetch maximum profits per unit investment.

15.6 RECOMMENDATIONS AND FUTURE LINE OF WORK

Drip irrigation at 80% CPE with 100% recommended dose of fertilizer (60: 30: 30 kg NPK ha^{-1}) as 75% conventional fertilizer + 25% water soluble fertilizer would be an optimum irrigation schedule for achieving higher level of productivity and economic return. Besides, this treatment also achieves water saving under castor + onion intercropping system. The results indicated the possibility of enhancing water and nutrient use efficiency resulting in higher yield of both castor and onion under drip fertigation. Hence, similar study may be extended to different agro-ecological zones to maximize the net return of the farm.

15.7 SUMMARY

Field experiments were conducted in farmer's holding at Kokalai village, Namakkal district during *Rabi* 2011–12 and 2012–13 to study the effect of drip irrigation regimes and fertigation levels on growth, yield, and economics of castor hybrid YRCH 1 + onion (variety CO 3) intercropping system. Experiments were laid out in split plot design during October 2011 – April 2012 and October 2012 – March 2013. Main plots were I_1 40% cumulative pan evaporation (CPE), I_2 –60% CPE, I_3 –80% CPE and fertilizer levels in subplots were: F_1: 75% RDF as conventional fertilizer (CF); F_2: 75% RDF as water-soluble fertilizer (WSF); F_3: 75% RDF (75% CF + 25% WSF); F_4: 100% RDF as conventional fertilizer (CF); F_5: 100% RDF as water-soluble fertilizer (WSF); F_6: 100% RDF (75% CF + 25% WSF); and surface irrigation with soil application of 100% RDF as CF was the control for comparison. The drip irrigation was scheduled once in three days based on the crop requirement of water (WRc) in the treatment schedule of 40, 60 and 80% CPE. Surface irrigation (control) was scheduled based on IW/CPE ratio of 0.75. RDF was supplied in the treatment schedule to meet out the crop demand in various stages under drip irrigation and surface irrigation system.

15.7.1   CASTOR CROP

Drip fertigation at 80% CPE with 100% RDF as WSF significantly delayed the days to 50% for flowering during 2011–12 and 2012–13, respectively than other treatment combinations. Drip irrigation regime at 80% CPE with 100% RDF as WSF as fertigation at 90 DAS recorded higher plant height during 2011–12 and 2012–13 than the other treatments. Combined application of 100% RDF as WSF with drip irrigation at 80% CPE recorded significantly higher DMP of castor at 45 and 90 DAS during 2011–12 and 2012–13, respectively compared to other drip fertigation and surface irrigation methods. Significantly maximum leaf area index of castor (2.08) was recorded under drip irrigation at 80% CPE with 100% RDF as WSF during 2011–12 and similar trend was recorded during the second year compared to other treatments. Significantly higher CGR was observed under drip irrigation at 80% CPE with 100% RDF as WSF at 45 –90 DAS during both years. Drip irrigation at 40% CPE with 75% RDF as CF combination registered maximum specific leaf weight during 2011–12 and same trend was recorded during second year.

The yield attributes (viz., length of primary spike, total number of capsule plant^{-1}, total number of spike plant^{-1} and 100 seed weight) were significantly higher under drip irrigation at 80% CPE and fertilizer levels at 100% RDF as WSF during both the years. Drip fertigation at 80% CPE with 100% RDF as WSF significantly recorded higher castor seed yield (2619 and 2490 kg/ha) followed by 80% CPE with 100% RDF (75% RDF as CF + 25% RDF as WSF) recorded castor seed yield of 2576 and 2279 kg/ha during both the years. Significantly higher biological yield was recorded under drip fertigation at 80% CPE with 100% RDF as WSF compared to other treatments during 2011–12 and 2012–13, respectively. Higher harvest index recorded under drip irrigation at 80% CPE, among different fertilizer levels, fertilizer at 100% RDF as WSF during both the years. Nutrient uptake (NPK) was found better under drip irrigation at 80% CPE with fertilizer application at 100% RDF as WSF at 45 and 90 DAS over other drip fertigation treatments during two years of study.

15.7.2 ONION AS INTERCROP

Drip irrigation at 80% CPE showed increased plant height at 45 DAS and at harvest stages, whereas maximum plant height was recorded at application of 100% RDF as WSF at two growth stages during both the years. At 45 DAS and at harvest stage, number of leaf sheaths significantly higher at 80% CPE in both the years. Significantly more number of leaf sheaths was recorded at 100% RDF as WSF in two year study period. Drip fertigation at 80% CPE with 100% RDF as WSF combination at 45 and 90 DAS recorded significantly higher DMP compared to other treatment combinations during both the years. Significantly higher CGR value at 45 –90 DAS was observed under drip irrigation at 80% CPE with fertilizer application at 100% RDF as WSF during 2011–12 and 2012–13, respectively over other treatment combinations. The yield contributing characters (viz., equatorial diameter, polar bulb diameter, number of bulb plant^{-1}, single bulb weight and bulb yield plant^{-1}) were significantly influenced by drip irrigation at 80% CPE, irrespective of fertilizer levels, and fertilizer at 100% RDF as WSF recorded similar yield attributes of onion intercrop during both the years. Drip fertigation at 80% CPE with 100% RDF as WSF significantly produced maximum bulb yield plant^{-1} (109.8 and 81.0 g) during both years.

Treatment combination of drip irrigation at 80% CPE with fertilizer application at 100% RDF as WSF significantly produced higher marketable onion bulb yield (5316 and 4577 kg/ha) during 2011–12 and 2012–13, respectively. Treatment combination of drip irrigation at 80% CPE with fertilizer application at 100% RDF as WSF significantly recorded maximum nutrient uptake (N, P, and K) at 45 DAS and at harvesting stage of onion intercrop over other fertigation combination during 2011–12 and 2012–13, respectively.

Drip fertigation at 80% CPE with fertilizer application at 100% RDF as WSF significantly recorded higher castor equivalent yield (6416 and 5759 kg/ha) compared to other treatment combinations during both the years. Application of 80% CPE with 100% RDF as WSF showed higher agronomic efficiency during 2011–12; whereas in 2012–13, 40% CPE with 75% RDF as WSF recorded higher AE of castor + onion intercrop.

In drip irrigation method, 80% CPE consumed a total amount of 388 and 456 mm of water during 2011–12 and 2012–13. In total, 560 and 539 mm of water was used under surface irrigation, in each year.

Drip irrigation at 40% CPE showed higher WUE than surface method of irrigation during 2011–12 and 2012–13, respectively. Maximum WUE was recorded in 100% RDF as WSF application during both the years. Significantly higher WUE was recorded under drip irrigation at 40% CPE with 100% RDF as WSF during 2011–12. At 80% CPE, water saving was 30.8% during 2011–12, and 15.5% in 2012–13 compared to surface method of irrigation.

Soil moisture content under surface irrigation method was steeply declined from 22.87% in top layer of 0 –15 cm depth at 2 days after irrigation (DAI) to 12.14% on 14 DAI. Under drip irrigation, the soil moisture content observed at 80% CPE always maintained above 85% available soil moisture even at 30 cm distance across the lateral at 15–30 cm depth for 24 and 48 hours after irrigation.

At the end of fertigation across the lateral, the peak available soil nitrogen significantly higher at 100% RDF as WSF in the depth of 15–30 cm at a distance of 15 and 30 cm from the dripper. With respect to lateral, it was steadily increased up to 20 cm distance from the dripper with 30 cm depth. The peak availability of phosphorus recorded just below the dripper. Across the lateral, higher available potassium recorded near the emitter point and decreased as the distance increases with lower level to 30 cm depth on distance and depth.

Drip irrigation at 80% CPE with 100% RDF as WSF combination recorded significantly maximum root weight and root volume of castor during both the years.

In onion intercrop, maximum root length was recorded at drip irrigation at 40% CPE, whereas application of 75% RDF as CF registered higher root length when compared to other treatments during both the years. Scheduling irrigation through drip at 80% CPE with fertigation at 100% RDF as WSF registered higher land equivalent ratio of 1.59 and 1.52 during 2011–12 and 2012–13 and superior over other treatment combinations.

Drip fertigation at 80% CPE with 100% RDF (75% CF + 25% WSF) had recorded mean net return of Rs. 1,63,921 per ha (2732 US$/ha) compared to 80% CPE with 100% RDF as water soluble fertilizer (Rs. 1,61,449 per ha (2691 US$/ha)) in 2011–12. During the year 2012–13, maximum net return recorded under drip fertigation at 80% with 100% RDF as WSF (Rs. 1,32,840 per ha (2214 US$/ha)) followed by 80% CPE with 100% RDF (75% CF + 25% WSF) (Rs. 1,31,223 per ha (2187 US$/ha)).

Drip irrigation at 80% CPE with 100% RDF as WSF resulted in higher net present value (Rs. 97,019 per ha (1617 US$/ha)) and discounted benefit-cost ratio (2.30), whereas under drip irrigation at 80% CPE with 100% RDF as (75% CF + 25% WSF) recorded net present value (Rs. 85014 per ha (1417 US$/ha)) and discounted benefit-cost ratio (2.14) during two year study.

KEYWORDS

- **benefit cost ratio**
- **castor**
- **CPE**
- **drip fertigation**
- **intercropping**
- **net present value**
- **onion**
- **RDF**
- **surface irrigation**
- **WSF**

REFERENCES

Abdullah, K., Tuylua, G. I., Ucarb, Y., & Cakmak, B., (2005). Crop water use of onion (*Allium cepa L)* in Turkey. *Agric. Water Manage., 72*, 59–68.

Aglave, B. N., Jagtap, P. K., & Mutkule, D. S., (2010). Economic analysis of different phosphorus levels on castor based intercropping system. *J. Oilseeds Res., 27*, 336–338.

Akpan, U. G., Jimoh, A., & Mohammed, A. D., (2006). Extraction, characterization and modification of castor seed oil. *Leonardo J. Sci., 5*(8), 43–52.

Al-Moshileh, A. M., (2001). Effect of nitrogen, phosphorus and potassium fertilizers on onion productivity in central region of Saudi Arabia. *Assiut. J. Agric. Sci., 32*, 291–305.

Alva, A. K., & Syvertsen, J. P., (1991). Irrigation water salinity affects soil nutrient distribution, root density and leaf nutrient levels of citrus under drip fertigation. *J. Plant Nut., 14*(7), 715–727.

Anilkumar, P., (2001). *Irrigation and Nitrogen Management Through Drip Fertigation in Chilli*. M.Sc (Ag).Thesis, Acharya N. G. Ranga Agricultural University (ANGRAU), Rajendra Nagar, Hyderabad, p. 131.

Anjanikumar, J., Pal, N., & Singh, N., (2000). Phosphorus uptake and utilization at crop growth stages by onion. In: *Proc. The National Symposium on Onion and Garlic and Postharvest Management Challenges and Strategies* (pp. 192–195). New Delhi.

Annandale, J. G., Jovanovic, N. Z., Campbel, G. S., Du Sautoy, N., & Lobit, P., (2004). Two dimensional solar radiation interpretation model for hedge row fruit trees. *Agril. Fore. Meteor., 121*(4), 207–225.

Anonymous, (1997). Agricultural productions summary, Turkish Republic. National Institute of Statistics, Ankara, p. 51.

Ansary, S. H., Choudhury, J., & Sarkar, S., (2006). Postharvest studies of onion (*Allium cepa*, L.) grown under different moisture regimes and fertilizer levels. *Crop Res., 31*(3), 404–407.

Arulkar, K. P., Sarode, S. C., & Bhuyar, R. C., (2008). Wetting pattern and salt distribution in drip and micro sprinkler irrigation. *Agric. Sci. Digest, 28*(2), 124–126.

Ashokkumar, R., Gautam, C., & Kaushik, S. K., (1995). Production potential of rain-fed pearl millet (*Pennisetum glaucum*) - castor (*Ricinus communis* L.) intercropping at different fertility levels. *Indian J. Agric. Sci., 65*(5), 315–322.

Ashokkumar, S., & Singh, K. G., (2006). Performance evaluation of drip irrigation system for different crop sequence. *J. Res. Punjab Agric. Univ., 43*(2), 130–133.

Aujla, M. S., Thind, H. S., & Buttar, G. S., (2005). Cotton yield and water use efficiency at various levels of water and N through drip irrigation under two methods of planting. *Agric. Water Mgt., 71,* 167–179.

Babita, M., Maheswari, M., Rao, L. M., Arun, K. S., & Gangadhar, R. D., (2010). Osmotic adjustment, drought tolerance and yield in castor (*Ricinus Communis* L.) hybrids. *Environ. and Exper. Bot., 69*(3), 243–249.

Bafna, A. M., Daftardar, S. Y., Khade, K. K., Patel, P. V., & Dhotre, R. S., (1993). Utilization of nitrogen and water by tomato under drip irrigation system. *J. Water Manage, 1,* 1–5.

Balasubramanyam, R., (2003). Drip irrigation and fertigation of onion. In: *Proc. Approaches for Sustainable Development of Onion and Garlic* (pp. 135–156). NHRDF, Nhasik.

Balasubramanyam, R., Dhake, A. V., & Moitre, P., (2001). Micro irrigation and fertigation of onion. In: *Proc. International Conference on Micro and Sprinkler Irrigation Systems* (Vol. 4, pp. 96–101). JISL.

Balasubramanyam, V. R., Dhake, A. V., & Moitra, P., (1999). Improvement of onion cultivation. *Agro India, 4,* 24–27.

Baldwin, B. S., & Cossar, R. D., (2009). Castor yield in response to planting date at four locations in the south central United States. *Ind. Crops Prod., 29,* 316–319.

Bandyopadhyay, K. K., Prakash, A. H., Sankaranarayanan, K., Dharajothi, B., & Gopalakrishnan, N., (2009). Effect of irrigation and nitrogen on soil moisture dynamics, productivity and input-use efficiency of Bt-cotton (*Gossypium hirsutam*) in a *Vertic ustropept. Indian J. Agri. Sci., 79*(6), 448–453.

Bangar, A. R., & Chaudhari, B. C., (2004). Nutrient mobility in soil uptake, quality and yield of *Suru* sugarcane as influenced by drip - fertigation in medium Vertisols. *J. Indian Soc. Soil. Sci., 52*(2), 164–171.

Barros-Junior, G., Guerra, O. C., Hugo, L. F., Cavalcanti, M., & De Lacerda, R. D., (2008). Water consumption and water use efficiency of two castor bean cultivars. *Revista Brasileira de Engenharia Agrícolae Ambiental., 12*(4), 350–355.

Barrs, H. D., & Weatherley, P. E., (1962). Re-examination of the relative turgidity technique for estimating deficit in leaves. *Aust. J. Biol. Sci., 15,* 413–428.

Bar-Yosef, B., & Sagiv, B., (1985). Potassium supply to field crops under drip irrigation and fertilization. In: *Proceedings of the K Symposium, International Potash Institute, Pretoria* (pp. 185–188).

Bar-Yosef, B., (1999). Advances in fertigation. *Adv. Agron., 65*, 1–77.

Basith, M. A., & Mohammad, S., (2010). Effect of intercropping on seed yield and economics of castor (*Ricinus communis* L.) grown under rain-fed conditions. *J. Oilseeds Res., 27*, 366–369.

Begum, R. A., Malik, S. A., & Rahman, M., (1990). Yield response of onion as influenced by different soil moisture regimes, *Bangladesh J. Agric. Res., 15*, 64–69.

Bekele, S., & Tilahum, K., (2007). Regulated deficit irrigation scheduling of onion in a semiarid region of Ethiopia. *Agric. Water Manage, 89*(1–2), 148–152.

Bharambe, P. R., Mungal, M. S., Shelke, D. K., Oza, S. R., Vaishnava, V. G., & Sondge, V. D., (2001). Effect of soil moisture regimes with drip on special distribution of moisture, salts, nutrient availability and water use in banana. *J. Indian Soc. Soil Sci., 49* (4), 658–665.

Bharambe, P. R., Narwade, S. K., Oza, S. R., Vaishnava, V. G., Shelke, D. K., & Jadhav, G. S., (1997). Nitrogen management in cotton through drip irrigation. *J. Indian Soc. Soil Sci., 45*(4), 705–709.

Bhonde, S. R., Singh, N. B., & Singh, D. K., (2003). Studies on the effect of drip irrigation in onion crop. *NHRDF News Letter, 13*(1), 1–2.

Billsegars, (2018). *Fertigation in IMC Global: Efficient Fertilizer Use Manual, 4ᵗʰ edn.,* 2003, http://www.agcentral.com/imcdemo/10fertigation.Html.Accessed on August 31.

Black, J. D. F., (1969). *Trickle Irrigation: A review. Hort. Abstr., 46*, 69–74.

Blaine, H., Hopmans, J. W., Simunek, J., & Gardenas, A., (2004). Crop nitrate availability and nitrate leaching under micro irrigation for different fertigation strategies. *Paper Presentation at Am. Soc. Agric. and Biol. Engineers.* Annual Meeting, St Joseph – MI, p. 6.

Bobade, S. S., (1999). *Studies on Drip Irrigation and Nitrogen Levels on the Water Use and Yield of Eggplant* (CO 2). M. Sc. (Ag.) Thesis, Tamil Nadu Agricultural University, Coimbatore, p. 122.

Bobade, S. S., Asokaraja, N., & Murali-Arthanari, P., (2002). Effect of drip irrigation and nitrogen levels on yield, water use efficiency of eggplant. *Crop Res., 24*(3), 481–489.

Bohm, W., Maduakor, H., & Taylor, H. M., (1977). Comparison of five methods for characterizing soybean rooting density and development. *Agronomic J., 69*, 415–419.

Bucks, D. A., & Davis, S., (1986). Historical development. In: Nakayama, F. S., & Bucks, D. A., (eds.), *Trickle Irrigation for Crop Production, Design, Operation and Management* (pp. 1–21). Elsevier Science Publisher, New York.

Bybordi, A., & Malakouti, M. J., (2003). The effect of various rates of potassium, zinc and copper on the yield and quality of onion under saline conditions in two major onion growing regions of east Azerbaijan. *Agric. Sci. and Tech., 17*, 43–52.

Carrijo, O. A., & Hochmuth, G., (2000). Tomato responses to pre-plant incorporated or fertigated phosphorus on soils. *Hort. Sci., 35*(1), 67–72.

Chakraborty, D., Singh, A. K., Kumar, A., & Khanna, M., (1999). Movement and distribution of water and nitrogen in soil as influenced by fertigation in Broccoli (*Brassica oleracea var. italica.* L.). *Agric. Water Manage, 7*(1 & 2), 8–13.

Chawla, J. K., & Narda, N. K., (2001). Economy in water and fertilizer use in trickle fertigated potato. *Irrig. and Drain., 50*, 129–137.

Clark, G. A., & Smajstrla, A. D., (1995). Comparative study of drip irrigation over furrow irrigation. In: *Micro Irrigation for a Changing World*. In: *Proc. of the 5th International Micro Irrigation Cong.* (pp. 526–531). Orlando, Florida, USA, ASABE, St Joseph – MI.

Corgan, J. N., & Kedar, N., (1990). Onion cultivation in subtropical climates. In: *Onion and allied Crops* (pp. 31–47). CRC Press, Boca Raton, Florida.

Corgan, J. N., Wall, M. M., Cramer, C. S., & Sammis, T., (2000). *Bulb Onion Culture and Management.* New Mexico Coop. Ext. Serv. Circ. 563, *Cooperative Agricultural Extension Service, University of New Mexico*, p. 11.

Dakshinamurthy, C., & Gupta, R. P., (1968). *Practices in Soil Physics*. IARI, New Delhi (Mimeographed), p. 58.

Damodaram, T., & Hedge, D. M., (2011). *Oilseeds Situations: A Statistical Compendium.* Directorate of Oilseeds Research Institute, Hyderabad, p. 53.

Dawelbeit, S., & Richter, C., (2004). Fertigation of onion crops by using surface irrigation in Sudan. In: *Proc. of Rural Poverty Reduction Through Research for Development* (pp. 23–34). Deutscher, Tropentag, Barlin.

Deek, I. M., Battikhi, A. M., & Khattari, S., (1997). Effect of irrigation and N fertilization (fertigation) scheduling on tomato in the Jordan valley. *J. Agron. and Crop Sci., 178*(4), 205–209.

Dhurandher, R., Pandey, N., & Tripathi, R. S., (1995). Effect of irrigation and nitrogen on dry matter yield and water requirement of summer soybean (*Glycine max* L.). *J. Oilseeds Res., 12*(2), 304–306.

Dimri, D. C., & Singh, V. P., (2005). Response of farm yard manure, nitrogen and row spacing on bulb weight and yield of onion (*Allium cepa* L.) Cv. VL –3. *Progressive Hort., 37*(1), 185–187.

Doan, L. G., (2004). Ricin: Mechanism of toxicity, clinical manifestations, and vaccine development. *A Review J. Tropical, 42*, 201–208.

DOR, (1993). *Research Highlights 1967–1992.* Directorate of Oilseeds Research Institute, Hyderabad, p. 45.

Drew, M. C., & Lynch, J. M., (1983). Soil microorganisms and root function. *Annual Rev. Phytopathology, 18*, 37–66.

El - Sherif, A. F., Shata, S. M., & Youssef, R. A., (1993). Effect of rates and method of zinc application on growth and nutrient uptake of tomato plants. *Egypt. J. Hort., 17*, 123–129.

Ennahli, S., & Earl, H. J., (2005). Physiological limitations to photosynthetic carbon assimilation in cotton under water stress. *Crop Sci., 45*, 2374–2382.

FAO: Crop Water in Cotton, (1977). http://www.fao.org/ag/ag/aglw/cropwater/cotton.stm Accessed on August 31, 2018.

Fontes, P. C. R., (1998). *Cultura da Cebola (Onion culture)*. Cadernos Didaticos 26, UFV, Vicosa - Brasil, p. 40.

Fontes, P. G. R., Sampaio, R. A., & Finger, F. L., (2000). Fruit size, mineral composition and quality of trickle irrigated tomatoes as affected by Potassium rates. *Pesquisa Agropecuaria Brasileira, 35*, 21–25.

Fujiyama, H., & Nagal, T., (1987). Studies on improvement of nutrient and water supply in crop cultivation on sand dune soil, II: Effect of fertilizer placement and irrigation method on growth and nutrient uptake of tomatoes. *Soil Sci. Pl. Nutr., 33*(3), 461–470.

Gala, P. T., (1992). Economics of tomato and capsicum production under drip versus furrow irrigation: Farmer's experience. In: *Proc. National Seminar on Drip irrigation, IPCL, Baroda* (pp. 69–76). Oxford and IBH Publ. Co., New Delhi.

Gangasaran, K., & Gajendra, G., (1983). Intercropping of dry land castor planted in different dates and planting systems with grain legumes. *Indian J. Agron., 28*(4), 362–328.

Giridhar, H., & Gajendra, G., (1991). Growth and yield of summer castor (*Ricinus communis* L.) and green gram (*Phaseolus radiatus*) intercropping. *Indian J. Agric, Sci., 61*(9), 669–671.

Glaser, L. K., Roetheli, J. C., Thompson, A. E., Brigham, R. D., & Carlson, K. D., (1993). Castor and *Lesquerella*: Sources of hydroxy fatty acids. In: *1992 Yearbook of Agriculture* (pp. 111–117). USDA Office Publishing Visual Communication, Washington – DC.

Grant, C. A., Flaten, D. N., Tamasiewicz, D. J., & Sheppard, S. C., (2001). Importance of early season phosphorous nutrition. *Can. J. Plant Sci., 81*, 211–224.

Hagin, J., & Lowengart, A., (1996). Fertigation for minimizing environmental pollution by fertilizers. *Fert. Res., 43*(1–3), 5–7.

Halvorson, A. D., Follet, R. F., Bartolo, M. E., & Schweissing, F. C., (2002). Nitrogen fertilizer use efficiency of furrow - irrigated onion and corn. *Agron. J., 94*, 442–449.

Hansen, W. R., (1972). Net photosynthesis and evapotranspiration of field grown soybean canopies. *PhD Thesis*, Iowa State University Library, Iowa, USA, p. 213.

Hanson, B. R., May, D. M., & Schwanki, L. J., (2003). Effect of irrigation frequency on subsurface drip irrigated vegetables. *Hort. Tech., 13*, 115–120.

Harjinder, S., Narda, N. K., & Chawla, J. K., (2004). Efficacy of phosphorus through trickle fertigation of potato (*Solanum tuberosum*). *Indian J. Agri. Sci., 74*(8), 476–478.

Hartz, T. K., & Hochmuth, G. J., (1996). Fertility management of drip irrigated vegetables. *Hort. Tech., 6*, 168–172.

Haynes, R. J., (1990). Movement and transformations of fertigated nitrogen below trickle emitters and their effects on pH in the wetted soil volume. *Fert. Res., 23*, 105–112.

Hearn, A. B., (1994). The principle of cotton water relations and their application in management. In: *Challenging the Future: Proc. of the World Cotton Research Conference I*, CSIRO, Australia, Brisbane, Australia, pp. 66–92.

Hebbar, S. S., Ramachandrapppa, B. K., Nanjappa, H. V., & Prabhakar, M., (2004). Studies on NPK drip fertigation in field grown tomato (*Lycopersicum esculantum* L.). *Eur. J. Agron., 21*, 117–127.

Heyman, D. S., & Mosse, B., (1972). Plant growth responses to VAM. *New Phytologist., 71*, 41.

Hochmuth, G. J., & Smajstrla, A. G., (2000). *Fertilizer Application and Management for Micro (Drip)- Irrigated Vegetables in Florida*. Florida co-operative extension service circular 14100, Univ., of Florida, http//www.Edis/ifas.ufl.edu/pdffiles/cv/cv14100.pdf. Accessed on August 31, 2018.

Hochmuth, G. J., (1992). Fertilizer management for drip-irrigated vegetables in Florida. *Hort. Tech., 2*(1), 27–32.

Howell, T. A., Bucks, D. A., & Chestness, J. L., (1981). Advances in trickle irrigation. In: *Proc. National Irrigation Symposium, Phoenix – AZ* (pp. 62–64).

Humphries, E. C., (1956). Mineral components and ash analysis. In: *Modern Methods of Plant Analysis* (pp. 468–502). Springer-Verlang, Berlin.

Itnal, C. J., Nagalikar, V. P., Lingaraju, B. S., & Basavaraj, P. K., (1994). Intercropping pigeonpea with pearl millet in Northeastern dry zone of Karnataka, *Karnataka J. Agril. Sci., 7*(1), 6–9.

Jackson, M. L., (1973). *Soil Chemical Analysis.* Prentice Hall of India Pvt. Ltd., New Delhi, p. 498.

Jacob, A., & Vexkull, H. V., (1958). *Fertilizer Use Nutrition and Manuring of Tropical Crops.* Velog for Ackerban, Hanover, W. Germany, p. 231.

Janagoudar, B. S., Venkatasubbiah, K., Janardhan, K. V., & Panchal, Y. C., (1983). Effect of short-term stress on free proline accumulation, relative water content and potassium content in different plant parts of three cotton genotypes. *Indian J. Plant Physiol., 26,* 82–87.

Janat, M., & Somi, G., (2001). Performance of cotton crop grown under surface irrigation and drip fertigation, II: Field water-use efficiency and dry matter distribution. *Commun. Soil Sci. Plant Anal., 32*(19 & 20), 3063–3076.

Janat, M., (2004). Assessment of nitrogen content, uptake, partitioning and recovering by cotton crop grown under surface irrigation and drip fertigation by using isotopic technique. *Commun. Soil Sci. Plant Anal., 35*(17 & 18), 2515–2535.

Jiang, Y., & Huang, B., (2000). Effects of drought or heat stress alone and in combination on Kentucky bluegrass. *Crop Sci., 40,* 1358–1362.

Kafkafi, U., Yosef, B., Rosenberg, R., & Sposito, G., (1988). Phosphorus adsorption by Kaolinite and montmorillonite: II. Organic anion competition. *Soil Sci. Soc. Amer. J., 52,* 1585–1589.

Kamalkumar V., Kendurkar, P. S., Madhuvajpeyi, A. S., & Singh, N., (2009). Oil quality characteristics and fatty acid composition of castor bean (*Ricinus communis* L.) cultivars. *J. Oilseeds Res., 26*(2), 183–186.

Keyvan, S., (2010). The effect of drought stress on yield, relative water content, proline, soluble carbohydrates and chlorophyll of bread wheat cultivars. *J. Anim. Plant Sci., 8*(3), 1051–1060.

Khepar, S. D., Neogy, P. K., & Kausal, M. P., (1983). Moisture and salt distribution pattern under drip irrigation. In: *Proc. Second National Seminar on Drip Irrigation* (pp. 34–40). TNAU, Coimbatore - 3.

Kittock, D. L., Williams, J. H., & Hanway, D. G., (1967). Castor bean yield and quality as influenced by irrigation schedules and fertilization rates. *Agron. J., 59,* 463–467.

Klepper, B., (1991). Crop root system responses to irrigation. *Irrig. Sci., 12,* 105–108.

Konton, R. A. L., Abbey, L., & Gbena, R. H., (2003). Irrigation schedule affects onion (*Allium cepa* L.) growth, development and yield. *J. Veg. Crop Prod., 9,* 3–11.

Konton, R. A. L., Abbey, L., & Gbena, R. H., (2004). Studies on water and potassium dynamics in soil under fertigation and furrow irrigation in radish. *J. Veg. Crop Prod., 6,* 44–51.

Kovach, S. P., Curtis, L. M., Tyson, T. W., & Hairston, J. E., (1999). Applying fertilizers through the system. In: *Micro Irrigation Handbook* (pp. 23–34). Alabama Irrigation and Water Resources Institute, College of Agriculture, Auburn University, Auburn, USA.

Kozhushka, L. F., & Romanets, V., (1994). Ecological and economic efficiency of mineral fertilizer application as a component of irrigation water. In: *17ᵗʰ ICID European Reg. Conf. Irrig. Drain.,* www.icid.org, 16–22 May, pp. 209–213.

Kreig, D. R., & Sung, F. J. M., (1986). Source - sink relationships as affected by water stress. In: Mauney, J. R., & Stewart, J. M., (eds.), *Cotton Physiology* (pp. 73–78). The Cotton Foundation, Memphis, Tenn.

Kumar, S., Sushant, C. P., Tiwari, C. P., & Singh, V., (2006). Bulb yield and quality of onion (*Allium cepa* L.) as affected by application rates of nitrogen and potassium fertilizer. *Agric. Sci. Digest., 26*(1), 11–14.

Kumavat, S. M., & Dhakar, L. L., (2000). Effect of irrigation regimes and nitrogen on yield and economics of soyabean. *J. Oilseeds Res., 17*(2), 295–298.

Lakkad, L. V., Asgdaria, K. B., Mathukia, R. K., Sagarka, B. K., & Khanpara, V. D., (2005). Efficient use of water and fertilizers through drip system for maximizing castor production. In: *Proc. National Conference on Sustainable Management of Water Resources* (pp. 146–147). New Delhi.

Lakshmamma, P., Lakshmi, P., Anjani, K., & Lavanya, C., (2010). Identification of castor genotypes for water uses efficiency (WUE) and root triats. *J. Oilseeds Res., 27*, 187–189.

Lamm, F. R., (2005). SDI for conserving water in corn production. In: Walton, R., (ed.), *Paper Presentation at World Water and Environmental Resources Congress* (p. 12). Anchorage, Alaska, USA.

Leelarani, P., (2008). Study on castor (*Ricinus communis* L.) based intercropping system under rain-fed conditions. *J. Oilseeds Res., 25*, 92–93.

Li, J., Zhang, J., & Ren, L., (2003). Water and nitrogen distribution as effected by fertigation of ammonium nitrate from a point source. *Irrig. Sci., 22*, 19 30.

Li, L., & Staden, V., (1998). Effects of plant growth regulators on the antioxidant system in seedlings of two maize cultivars subjected to water stress. *Plant Growth Regulation, 25*, 81–87.

Locascio, S. J., Olson, S. M., & Rhoads, F. M., (1989). Water quantity and time of N and K application for trickle-irrigated tomatoes. *J. Amer. Soc. Hort. Sci., 114*, 265–268.

Machado, R. M. A., Oliveira, M. R. G., & Portas, C. A. M., (2003). Tomato root distribution, yield and fruit quality under subsurface drip Irrigation. *Plant and Soil, 255*, 333–341.

Madhusudana-Rao, L., & Venkateswarlu, M. S., (1988). Effect of plant densities, irrigation and graded dose of N on oil content and nutrient uptake of castor. *J. Res. APAU, 16*, 43–46.

Malavia, D. D., & Khanpara, V. D., (1995). Comparison of irrigation methods in arid and semiarid western Gujarat. In: *Proceedings of 5ᵗʰ International Micro irrigation Congress* (pp. 464–469). Orlanda, Florida, USA, ASABE, St Joseph – MI, 2–6.

Martin, G. C., Fergusson, L., & Palito, V. S., (1994). Flowering, pollination, fruiting, alternate bearing and abscission. In: Ferguson, L., Sibbett, G. C., & Martin, G. C., (eds.), *Olive Production Manual, No. 3354*. Division of Agriculture and Natural Resources, University of California, California.

Mathukia, R. K., & Modhwadia, M. M., (1993). Response of castor (*Ricinus communis* L.) to nitrogen and phosphorus. *Indian J. Agron., 38*(1), 152–153.

Mathukia, R. K., & Modhwadia, M. M., (1995). Influence of different levels of nitrogen and phosphorus on yield and nutrient uptake by castor (*Ricinus communis* L.). *GAU Res. J., 21*(1), 149–151.

Maya, P., (1996). *Studies on the Effect of Spacing Cum Nitrogen and Phosphorus on Growth, Yield and Quality of Sweet Pepper*. Cv. California Wonder. M.Sc. (Ag.) Thesis, Tamil Nadu Agricultural University, Coimbatore, p. 143.

Meenakshi, N., & Vadivel, E., (2003). *Growth and Productivity of Hybrid Bitter Guard (Homoridica charantia L.)* COB GOH −1 under different macro- and micro- nutrient fertigation levels. PhD (Hort.) Thesis, Tamil Nadu Agricultural University, Coimbatore, pages 288.

Michelakhis, N., Vougioucalou, E., & Clapaki, G., (1993). Water use, wetted soil volume, root distribution and yield of avocado under drip irrigation. *Agric. Water Mgt., 24,* 119–131.

Mingochi, D. S., (1998). FAO special program for food security in Zambia: Small holder irrigation and water program - irrigation component. *Food and Agric. Org.,* Rome, pp. 123–128.

Mishra, K. K., & Pyasi, S. K., (1993). Moisture distribution pattern in drip irrigation. *J. Res. Birsa Agric. Univ., 5*(1), 59–64.

Mnolawa, K. B., (2008). Potassium nitrate dynamics under drip irrigation for cropped and non-cropped conditions. *Botswana J. Agric. and Applied Sci., 54,* 25–29.

Mohammad, M. J., (2004). Utilization of applied nitrogen and irrigation water by drip fertigated squash as determined by nuclear and traditional techniques. *Nutr. Cycl. Agroecosyst., 68,* 1–11.

Mohammadkhani, N., & Heidari, R., (2008). Effects of drought stress on soluble proteins in two maize varieties. *Turk. J. Biol., 32,* 23–30.

Moreira, L. G., Viana, T. V. de A., Marinho, A. B., Nobre, J. G. A., Lima, A. D., & Albuuquerque, A. H. P., (2009). Efeito de diferntes laminas de irrigacao na produtividade da mamoneira variedade IAC Guarani (Effect of different irrigation levels on the productivity of the castor variety IAC Guarani). *Revista Brasileira de Ciencias Agrarias., 4,* 449–455.

Naghabhushanam, U., & Raghavaiah, C. V., (2005). Seeding date and irrigation effects on the productivity and oil quality post monsoon grown castor. *J. Oilseeds Res., 22*(1), 206–208.

Nakayama, F. S., & Bucks, D. S., (1991). Water quality in drip / trickle irrigation - A review. *Irrig. Sci., 12,* 187–192.

Narayan, S. M., Prabhakara-Setty, T. K., & Sridhara, S., (2009*).* Performance of castor, *Ricinus communis* L. genotypes at different integrated nutrient management practices under irrigated conditions. *J. Oilseeds Res., 26*(1), 41–43.

Nazirbay, I., Evett, S. R., Esanbekov, Y., & Kamilov, B., (2005). Water use of maize for two irrigation methods and two scheduling methods. *Agronomy Abstracts, ASA-CSSA-SSSA Annual Meeting,* Salt Lake City, Utah, p. 88.

NRCOG, (2018). Research activities, 2002, http//nrcog.Mah.nic.in.Accessed on August 31.

Ober, E. S., Bloa, M. L., Clark, C. J. A., Royal, A., Jaggard, K. W., & Pidegon, J. D., (2005). Evaluation of physiological traits as indirect selection criteria for drought tolerance in sugar beet. *Field Crop Res., 91,* 231–249.

Obreza, T. A., & Vavrina, C. S., (1995). Fertilization scheduling for improved nutrient use efficiency of micro-irrigated bell pepper in sandy soil. In: *Proc. Dahha Greidinger International Symposium on Fertigation* (pp. 247–256). Israel.

Omprakash, K., & Bushan, L. S., (2000). Productivity and economics of pigeonpea (*Cajanus cajan*) and castor (*Ricinus communis*) based intercropping systems. *Indian J. Soil Conserv., 28*(2), 147–150.

Osava, M., (2003). *Energy in Castor Bean,* http:// www.tierramerica.net/english/ 2003 /0526/ianalisis.shtml Accessed on August 31, 2018.

Padmavathi, P., & Ragavaiah, C. V., (2004). Productivity and returns of castor (*Ricinus communis* L.) - based intercropping systems with pulses and vegetable under rain-fed conditions. *Indian J. of Agric. Sci., 75*(5), 235–238.

Pahalwan, D. K., & Tripathi, R. S., (1984). Irrigation scheduling based on evaporation and crop water requirement for summer peanuts. *Peanut Sci., 11,* 4–6.

Paida, V. J., (1976). *Effect of Nitrogen and Phosphorus Fertilization on Growth, Yield and Oil Content of Castor.* M.Sc. (Ag.) Thesis, Gujarat Agricultural University, Sardar Krushinagar, Anand, India, p. 110.

Palaniappan, S. P., (1985). *Cropping System in the Tropics: Principles and Management.* Wiley Eastern Ltd., New Delhi, p. 215.

Palanisami, K., Paramasivam, P., & Ranganathan, C. R., (2002). *Agricultural Production Economics. Analytical Methods and Applications.* Associated Publishing Company, New Delhi, p. 219.

Pandey, R. P., Solanki, P. N., Saraf, R. K., & Parihar, M. S., (1996). Effect of nitrogen and phosphorus on growth and yield of tomato varieties. *Punjab Veg. Growers, 31,* 1–5.

Pandey, V. K., Bharat, J. P., & Harshana, D. B., (2009). Water requirement of bitter gourd under pressurized irrigation system. *J. of Indian Water Resource Soc., 28*(2), 15–19.

Pandita, M. L., (2000). Current status of research and development in crop production techniques in onion and garlic and future strategies to meet the challenges of new millennium. *Paper Presented at the National Symposium on Onion - Garlic Production and Postharvest Management Challenges & Strategies,* Nasik, p. 7.

Panse, V. G., & Sukhatme, P. V., (1978). *Statistical Methods for Agricultural Workers, 2nd edn.,* ICAR, New Delhi, India, p. 214.

Papadopoulos, I., (1992). Fertigation of vegetables in plastic house: Present situation and future aspects. *Acta Hort., 323,* 151–174.

Patel, G. N., Patel, P. T., & Patel, P. H., (2008). Yield, water use efficiency and moisture extraction pattern of summer groundnut as influenced by irrigation schedules, sulfur levels and sources. *J. SAT Agric. Res.* <ejournal.icrisat.org>, *6,* 1–5.

Patel, K. S., Patel, G. N., & Patel, M. K., (2005). Nitrogen requirement of Rabicastor (*Ricinus communis* L.) under different crop sequences. *J. Oilseeds Res., 22*(1), 209–210.

Patel, K. S., Patel, M. K., Patel, G. K., & Pathak, H. C., (2007). Intercropping in castor (*Ricinus communis* L.) under irrigated conditions. *J. Oilseeds Res., 24*(1), 121–123.

Patel, K. S., Patel, M. K., Patel, G. N., & Pathak, H. C., (2006). Fertigation study in castor, *Ricinus communis* L. *J. Oilseeds Res., 23*(1), 125–128.

Patel, M. K., Fatteh, U. G., & Patel, V. J., (1991). Effect of spacing and sowing time on the yield of irrigated castor under North Gujarat conditions. *G.A.U. Res., J., 17*(1), 119–121.

Patel, N., & Rajput, T. B. S., (2002). Yield response of some vegetable crops to different levels of fertigation. In: *Proceedings of Agriculture in Changing Global Scenario, Indian Society of Agricultural Sciences,* IARI, New Delhi, pp. 21–23.

Patel, P. S., & Jaimini, S. N., (1991). Inter-relationship and path analysis of certain quantitative characters in castor (*Ricinus communis* L.). *J. Oilseeds Res., 8,* 105–108.

Patel, R. A., Patel, J. J., & Patel, A. S., (2010). Seed yield and net returns of drip irrigated late *kharif* castor (*Ricinus communis* L.) as influenced by plant geometry and nitrogen levels. *Internat. J. Agric. Sci., 6*(2), 449–452.

Patel, R. M., & Patel, B. K., (2012). Effect of fertility levels on yield of different castor (*Ricinus communis* L.) genotypes under irrigated conditions. *J. Oilseeds Res., 29*(2), 129–130.

Patel, R. M., Patel, M. M., & Patel, G. N., (2009). Effect of spacing and nitrogen levels on *Rabi*castor, *Ricinus communis* L. grown under different cropping sequences in North Gujarat agro - climatic conditions. *J. Oilseeds Res., 26*(2), 123–125.

Patil, M. V., (1999). *Comparison of Drip Along With Fertigation Versus Surface Methods of Irrigation With Solid Fertilizers for Eggplant.* M.Sc. (Ag.) Thesis, MPKV, Rahuri, MS, p. 131.

Patil, R. S., More, S. L., Firee, N. N., Desale, S. B., & More, T. A., (2000). Effect of micro irrigation system and nitrogen fertigation levels on yield and quality of white onion (cv. Phule Safed) during summer onion. In: *Presented at National Symposium on Onion and Garlic Production and Postharvest Management Challenges & Strategies*, Indian Soc. of Veg. Sci., New Delhi, p. 204.

Patil, S. J., Palakondareddy, P., Kenchanagoundar, P. V., & Patil, S. A., (1989). Method of estimation of leaf area in castor (*Ricinus communis* L.). University of Agricultural Sciences. *J. Oilseeds Res., 6*, 384–385.

Patil, V. S., & Janawade, A. D., (1999). Soil water and plant atmosphere relationships. In: *Proc. Advances in Micro Irrigation and Fertigation* (pp. 19–32). Dharwad, Karnataka.

Patil, V. T., (2005). Onion development in Maharashtra state and program for increasing production. In: *Presented at National Symposium on Current Trends in Onion, Garlic, Chilli and Seed & Spices Production, Marketing and Utilization* (pp. 145–151). NRCOG, New Delhi.

Pettigrew, W. T., (2004). Physiological consequences of moisture deficit stress in cotton. *Crop Sci., 44*, 1265–1272.

Piper, C. S., (1966). *Soil and Plant Analysis.* Inter Science Publisher Inc., New York, p. 431.

Pooran, C., & Sujatha, M., (2000). Castor based intercropping systems – A review. *Agric. Rev., 21*(4), 244–248.

Prabhakara, B. N., Ramachandrappa, B. K., Nanjappa, H. V., & Soumya, T. M., (2010). Effect of frequency and methods of fertigation on growth, yield, quality and economics of green chilli (*Capsicum annum* L.). *Mysore J. Agric. Sci., 44*(3), 523–528.

Pratap, A. K. R., Reddy, A., & Padmavathi, P., (2006). Effect of irrigation and integrated nutrient management on seed and oil yield of *Rabi* castor (*Ricinus communis* L.). *J. Oilseeds Res., 23*(2), 230–241.

Praveenrao, V., & Raikhelkar, S. V., (1993). Effect of irrigation and fertilization on growth and yield in sesame. *J. Oilseeds Res., 10*(1), 31–36.

Prince, C. A., Sanders, D. C., & Campbell, C. R., (1998). Response of bell pepper to N fertilizer and N/K ratios. In: *Proc. of Fourth International Micro Irrigation Congress in Australia* (pp. 7–10). ICID, New Delhi.

Purshotam, R., Singh, J. P., & Singh, B. P., (1989). Studies on row spacing and planting pattern in castor and green gram intercropping system. *Indian J. of Dryland Agric. Res. and Dev., 4*, 103–106.

Radford, P. J., (1967). Growth analysis formulae: Their use and abuse. *Crop Sci., 8*, 171–175.

Raghavaiah, C. V., (1999). Performance of castor under different levels of fertilizer in rain-fed conditions of Alfisols. *J. Oilseeds Res., 16*, 295–298.

Raj, A. D., Patel, B. S., & Mehta, R. S., (2010). Effect of irrigation methods on growth and economics of hybrid varieties of castor. *Indian J. Agric. Sci., 80*, 795–800.

Rajput, R. L., & Mishra, M. K., (1995). Studies on intercropping in castor (*Ricinus communis* L.) under rain-fed conditions. *Haryana J. Agron., 11*(2), 141–144.

Rajput, R. L., & Srivastava, U. K., (1996). Performance of castor based intercropping under rain-fed conditions. *Indian J. Agron., 42*(4), 550–552.

Rajput, T. B. S., & Neelam, P., (2006). Water and nitrate movement in onion under drip fertigation and irrigation treatments. *Agric. Water Manage, 45*, 454–467.

Ramanjaneyulu, A. V., Ramprakash, T., & Vishnuvardhanreddy, A., (2010). Effect of sowing date and irrigation scheduling on yield and economics of winter castor in southern Telangana zone of Andhra Pradesh. *J. Oilseeds Res., 27*, 180–181.

Rana, D. S., Giri, G., & Pachauri, D. K., (2006). Evaluation of castor (*Ricinus communis* L.) genotypes for productivity, economics, litter fall and changes in soil properties under different levels of inter-row spacing and nitrogen. *Indian J. Agron., 51*(4), 318–322.

Randhawa, N. S., & Venkateswaralu, J., (1980). Indian experience in semiarid tropics: Prospect and retrospect. In: *Proc. of the International Symposium on Development and Transfer of Technology for Rain-fed Agriculture and Farmers* (pp. 207–220). ICRISAT, Hyderabad, A. P., India.

Rani, S., (2006). Agronomic options to maximize the yield of baby corn (*Zea mays* L.) in periyar Vaigai Command area. M.Sc. (Ag.) Thesis, TNAU, Coimbatore, p. 110.

Rao, C. M., & Venkateswarlu, M. S., (1988). Effect of irrigation, nitrogen and plant density on yield attributes and yield of castor varieties. *J. of Res., APAU, 16*(1), 37–39.

Rao, D. S., Poonia, B. L., & Ahuja, D. B., (1995). Performance of castor (*Ricinus communis* L.) cultivars and their response to fertilizer and plant protection. *Indian J. Agric. Sci., 65*(9), 648–651.

Raushhkolb, R. S., & Rolston, D. E., (1978). Phosphorus fertilization with drip irrigation. *Soil Sci. Soc. Am., 40*, 68–72.

Ravi, B., Sujatha, S., & Balasimha, D., (2007). Impact on drip fertigation on productivity of arecanut (*Areca catechu* L.). *Agric. Water Manage, 90*, 101–111.

Reddy, C., & Reddy, M., (2005). Different levels of vermi-compost and nitrogen on growth and yield of onion (*Allium cepa* L.), radish (*Raphanus sativus* L.) cropping system. *J. Res., ANGRAU, 33*(1), 11–17.

Reddy, D. V. V., (1994). *Annual Report*. ANGRAU, Hyderabad (1993–1994). ANGRAU, Hyderabad, p. 92.

Renault, D., & Wallender, W. W., (2000). Nutritional water productivity and diets. *Agric. Water Manage., 45*, 275–296.

Richard, L. A., (1947). Pressure membrane apparatus, construction and use. *Agric. Eng., 28*, 451–454.

Rizauddin, A. S., Khadke, S. M., Chandrasekhar-Reddy, K., & Maruthi-Sankar, G. R., (2001). Soil test based optimal fertilizer requirements for attaining different yield targets of castor (*Ricinus communis* L.) in dry land Alfisols. *Indian J. Agric. Sci., 71*(1), 27–30.

Ruiz, J. M., & Romero, L., (1998). Commercial yield and quality of fruits of cucumber plants cultivated under greenhouse conditions: Response to increases in nitrogen fertilization. *J. Agric. Food Chem., 46*, 4171–4173.

Ryan, I., (2002). Efficient use of phosphate fertilizers for sustainable crop production in Wana. *IMPHOS: Phosphate Newsletter*, 2–5.

Sabale, R. N., & Khuspe, V. S., (1986). Irrigation studies in summer groundnut. *J. Maharashtra Agric. Univ., 11*, 9–13.

Salo, T., Suojala, T., & Kallela, M., (2002). The effect of fertigation on yield and nutrient uptake of cabbage, carrot and onion. In: *Proc. of Workshop Towards and Ecologically Sound Fertilization in Field Vegetable Production* (p. 571). Int. Soc. of Hort. Society (ISHS), Atlanta – GS, *Acta Hort.*

Sampathkumar, T., & Pandian, B. J., (2011). Nutrient uptake rate and relationship with grain yields of hybrid maize under drip fertigation. *Madras Agric. J., 98*(4–6), 151–153.

Sampathkumar, T., Krishnasamy, S., Ramesh, K., & Shanmugasundaram, K., (2006). Effect of drip and surface irrigation methods with rice straw mulch on productivity and water use efficiency of summer cotton. *Crop Res., 32*(2), 141–144.

Sankar, V., Lawendae, K. E., Qureshi, A. A., & Tripathi, P. C., (2005). Effect of bio power force application on growth and yield of onion. *The Orissa J. Agric., 33*(1), 4–7.

Sarma, D. A., (1985). Effect of varieties and levels of nitrogen on the yield of castor. *Andhra Pradesh J., 32*(2), 133–134.

Satisha, G. C., (1997). Fertigation - new concept in Indian agriculture. *Kisan World*, 29–30.

Satyandrakumar, M., Imtiyaz, P., & Kumar, A., (2007). Effect of differential soil moisture and nutrient regimes on postharvest attributes of onion (*Allium cepa* L.). *Sci. Horti., 112*, 121–129.

Sausen, T. L., & Goncalves-Rosa, L. M., (2010). Growth and carbon assimilation limitations in *Ricinus communis* L. under soil water stress conditions. *Acta Bot. Bras., 24*(3), 648–654.

Savitha, B. K., Paramaguru, P., & Pugalendhi, L., (2010). Effect of drip fertigation on growth and yield of onion. *Indian J. Hort., 67*, 334–336.

Segel, E., Bengal, A., & Shani, U., (2000). Water availability and yield response to high frequency micro irrigation in sunflower. In: *Proc. of Sixth National Microirrigation Congress on Microirrigation Technology for Developing Agriculture* (p. 324). South Africa.

Selvarani, A., (2009). Drip fertigation studies in maize (*Zea mays L.*) –Okra (*Abelmoschusb esculentus* L. Moench) cropping sequence. *PhD Thesis*, Tamil Nadu Agricultural University, Coimbatore, p. 311.

Senthilkumar, A., (2000). Effect of microsprinkler irrigation and fertigation on yield and quality of groundnut. *M.Sc.(Ag.) Thesis*, Tamil Nadu Agric. Univ., Coimbatore, p. 128.

Sesha-Saile, S. P., & Bhaskar-Reddy, B., (2005). Effect of tillage and soil moisture regimes on seedling emergence, growth and yield of summer castor in rice fallows. *J. Oilseeds Res., 22*(2), 327–330.

Severino, L. S., Auld, D. L., Marco, B., & Magno, C. G. C., (2012). Review on the challenges for increased production of castor. *Agron. J., 104*(4), 853–880.

Shadbolt, C. A., & Holm, L. G., (1956). Some quantitative aspects of weed competition in vegetable crops. *Weeds, 4*(2), 111–123.

Shamima, N., Imamul-Haq, S. M., & Hossain, M. A., (2003). Sulphur effects on growth responses and yield of onion. *Asian J. Pl. Sci., 2*(12), 897–902.

Sharathkumar, H. C., Mudulagiriyappa, H., Nanjappa, V., & Ramachandrappa, B. K., (2010). Productive performance of castor (*Ricinus communis* L.) based intercropping systems under rain-fed conditions of central dry zone in Karnataka. *Mysore J. Agric. Sci., 44*(33), 481–484.

Shinde, S. H., Bhoi, P. G., Raskar, B. S., Pawar, D. D., & Bangar, A. R., (2000). Effect of water soluble fertilizer applied through drip on growth and yield of cotton. In: *Proc. International Congress on Micro- and Sprinkler Irrigation System* (pp. 82–86). Jain Irrigation, Jalgaon, Maharashtra.

Simsek, M., Tonkaz, T., Kacira, M., Comlekcioglu, N., & Dogan, Z., (2005). The effects of different irrigation regimes on cucumber (*Cucumbis sativus* L.) yield and yield characteristics under open field conditions. *Agric. Water Mgt., 73,* 173–191.

Singandhupe, R. B., Rao, G. G. S. N., Patil, N. G., & Brahmanand, P. S., (2003). Fertigation studies and irrigation scheduling in drip irrigation system in tomato (*Lycopersicom esculentum* L.). *Euro. J. Agron., 19*(2), 327–340.

Singh, A. K., Chakraborty, D., Mishra, P., & Singh, D. K., (2002). Nitrogen and Potassium dynamics in fertigation systems. *Paper Presentation at 17th WCSS,* Bangkok – Thailand, p. 6.

Singh, D., & Sharma, R. P., (1991). Effect of moisture regimes and nitrogen fertilization of onion. *Indian J. Agron., 36*(1), 125–126.

Singh, I., (2009). Study of intercropping of castor, *Ricinus communis* L. under irrigated conditions. *J. Oilseeds Res., 26*(2), 170–171.

Singh, J. P., & Singh, S. P., (1988). Intercropping of mung bean and guar in castor under rain-fed conditions. *Indian J. Agron., 33*(2), 177–180.

Singh, P., Kumar, R., Agrawal, M. C., & Magal, J. L., (1990). Performance of drip and surface irrigation for tomato in heavy soils, In: *Proc. XI International Congress on the Use of Plastics in Agriculture* (pp. 51–56). IARI, New Delhi.

Singh, R. P., Jain, N. K., & Poonia, B. L., (2000). Response of *Kharif* onion to nitrogen, phosphorus and potash in eastern plains of Rajasthan. *Indian J. Agric. Res., 70,* 871–872.

Singh, R. R., & Ramakrishna, Y. S., (1975). Moisture use efficiency of dry land crops as influenced by fertilizer use in oilseed crops. *Annals of Arid Zone, 14*(4), 320–328.

Singh, R. R., Sharma, C., & Singh, B. P., (2003). Effect of NPK fertigation and planting patterns on yield and economics of potato under drip irrigation. *J. Indian Potato. Assoc., 30*(1–2), 69–70.

Singh, S. P., & Verma, A. B., (2001). Response of onion (*Allium cepa* L.) to potassium application. *Indian J. Agron., 46,* 182–815.

Sivanappan, R. K., & Padmakumari, O., (1980). *Drip Irrigation.* Bulletin No. 7, Tamil Nadu Agric. University, Coimbatore, p. 70.

Sivanappan, R. K., (1998). Status, scope and future prospects of micro irrigation and sprinkler irrigation systems. In: *Paper Presentation at Workshop on Micro Irrigation and Sprinkler Irrigation Systems* (p. 11). IARI, New Delhi.

Solaimalai, A., Baskar, M., Sadasakthi, M., & Subburamu, A., (2005). Fertigation in high value crops. A review. *Agric. Rev., 26*(1), 1–13.

Song, F. B., Dai, J. Y., Gu, W. B., & Li, H. Y., (1995). Effect of water stress on leaf water status in maize. *J. of Jilin Agric. Univ., 18,* 1–6 (in Chinese).

Sorensen, R. B., & Butts, C. L., (2005). Cotton, corn, and peanut yield under subsurface drip irrigation. In: Walton, R., (ed.), *World Water and Environmental Resources Congress at Anchorage* (p. 12). Alaska, USA.

Souza, A. S., Tavora, F. J. A. F., Pitombeiria, J. B., & Bezerra, F. M. I., (2007). Planting time and irrigation management for castor plant and its effect on growth and productivity. *Rev. Cienc. Agronomica, 38,* 422–429. (In Portuguese and English abstract).

Sree, P. S. S., & Reddy, B. R., (2003). Performance of castor cultivars at different dates of sowing. *Annals of Agri. Res., 24*(3), 546–551.

Srilatha, A. N., & Masthan, S. C., (2001). Evaluation of biological and economic efficiency in castor - legume intercropping system. *Res. Crops, 2*(3), 445–448.

Srilatha, A. N., Masthan, S. C., & Mohammed, S., (2002). Production potential of castor intercropping with legumes under rain-fed conditions. *J. Oilseeds Res., 19*(1), 127–128.

Srinivasa-Rao, M., Rama-Rao, C. A., & Ramakrishna, Y. S., (2006). *Crop – Crop Diversity as a Key Component of IPM in Pigeonpea.* Research Bulletin (CRIDA - ICAR), Hyderabad, p. 24.

Sririsha, A., Pradapkumar-Reddy, A., Padmavathi, P., & Madhu-Bindhu, G. S., (2010). Evaluation of different integrated nutrient management options on growth and yield of castor, Ricinus communis. L. *J. Oilseeds Res., 27*(1), 36–38.

Srivastava, S. K., & Chandra, D. R., (2009). Nutrient requirements of *kharif* castor, *Ricinus communis,* L. under irrigated conditions of Uttar Pradesh. *J. Oilseeds Res., 26*(2), 131–133.

Subbareddy G., Gangadhar-Rao, D., Venkateswaralu, S., & Maruthi, V., (1996). Drought management options for rain-fed castor in alfisols. *J. Oilseeds Res., 13*(2), 200–207.

Subbareddy, G., Maruthi, V., Vanaja, M., & Sree-Rekha, M., (2004). Influence of soil depth on productivity of castor and clusterbean in sole and intercropping systems. *Indian J. Agric. Sci., 38*(2), 79–86.

Subramanian, S., Sundar-Singh, S. D., Ramaswamy, K. P., Packiaraj, S. P., & Rajagopalan, K., (1974). Effect of moisture stress at different growth stages of groundnut. *Madras Agric. J., 61*, 813–814.

Sudhakar, C., & Praveen-Rao, V., (1998). Performance of different crops during post rainy season under varied moisture regimes in Southern Telangana region. *J. Res. A. P. A. U.,* 113–115.

Sudharani, C., (2000). Crop growth and development of castor cultivars under optimal and sob-optimal water and nitrogen conditions in Telangana region. *PhD Thesis,* Acharya N. G. Ranga Agricultural University (APAAU), Hyderabad, p. 278.

Sudharani, C., Buchareddy, B., & Myers, R. J. K., (2009). Effect of water regimes and nitrogen levels on growth degree days radiation use efficiency of castor cultivars. *J. Oilseeds Res., 26*, 363–365.

Suganya, S., Anitha, A., & Appavu, K., (2007). Moisture and nutrient distribution system under drip fertigation systems. In: *Third International Ground Water Conference on Water* (p. 12). Environment and agriculture – present problems and future challenges, IARI, New Delhi.

Tandon, H. L. S., (1991). *Fertilizer Management in Dry Land Agriculture.* FDCO, New Delhi, p. 81.

Thadoda N. K., Sukhadia, N. M., Malavia, D., & Moradia, A. M., (1996). Response of castor *(Ricinus communis L.)* GCH- 4 to planting geometry and nitrogen fertilization under rain-fed condition. *GAU. Res., J., 21*(2), 85–87.

Thompson, T. L., Thomas, A. D., & Ronald, E. G., (2000). Nitrogen and water Interactions in subsurface drip - irrigated cauliflower, part II: Agronomic, economic and environmental outcomes. *Soil Sci. Soc. Amer. J., 64*(1), 412–418.

Thompson, T. L., White, S. A., James, W., & Sower, G. J., (2003). Fertigation frequency for subsurface drip-irrigated broccoli. *Soil Sci. Soc. Am. J., 67*, 910–918.

Tumbare, A. D., & Bhoite, S. V., (2002). Effect of solid soluble fertilizer applied through fertigation on growth and yield of chilli. *Indian J. Agric. Sci., 72*(2), 109–111.

Tumbare, A. D., & Nikam, D. R., (2004). Effect of planting and fertigation on growth and yield of green chilli. *Indian J. Agric. Sci., 74*, 242–245.

Tumbare, A. D., Shinde, B. N., & Bhoite, B. U., (1999). Effect of liquid fertilizers through drip irrigation on growth of yield of okra. *Indian J. Agron., 44*(1), 176–178.

Tyler, K. B., & Lorenz, O. A., (1991). Fertilizer guide for California vegetable crops. Univ. Calif. Davis, Dept. *Veg. Crops,* (Special Publ.), p. 34.

Veeranna, H. K., Al-Khalak, P., & Sujith, G. M., (2000). Effect of fertigation and irrigation methods on yield, water and fertilizer use efficiencies in chilli (*Capsicum annuum* L.). *South Indian Hort., 49*, 101–103.

Venkateshwarlyu, S., & Reddy, G. S., (1989). Effect of time of planting of component crops on productivity of castor + cluster bean intercrops system. *J. Oilseeds Res., 6*, 308–315.

Venkateswarlu, S., (1986). Intercropping of pigeonpea with short duration pulses in semiarid Alfisols. *Indian J. Dry land Agric. Res. Develop., 1*, 48–55.

Verma, A. K., Kumar, S., & Tripathi, V., (2010). Optimization of fertilizer requirement for castor (*Ricinus communis* L.) under rain-fed conditions of Chattisgarh. *J. Oilseeds Res., 27*, 362–364.

Vignolo, R., & Naughton, F., (1991). Castor: A new sense of direction. *Int. News Fats Oils Relat. Mater., 2*, 692–699.

Vijaykumar, B., & Shivashankar, M., (1992). Response of castor (*Ricinus communis L.)* varieties to irrigation, nitrogen and plant density. *J. Res. APAU, 20*(1), 25–26.

Vijaykumar, B., (1992). Effect of irrigation, nitrogen and plant density on yield attributes and yield of castor varieties. *Indian J. Agron., 37*(1), 203–205.

Vishalakshi, K. P., Mathew, R., Suseela, P., & Bridgit, T. K., (2007). Soil moisture distribution pattern under surface and subsurface drip irrigation. In: *Paper Presentation at Third International Ground Water Conference on Water, Environment and Agriculture – Present Problems and Future Challenges* (p. 8). IARI, New Delhi.

Vishnumurthy, T., (1988). Contribution of production factors in castor on farmer's field. *Indian J. Dry land Agric. Res. Dev., 3*(1), 83–87.

Viswanathan, G. B., Ramachandrappa, B. K., & Nanjappa, H. V., (2002). Soil-plant water status and yield of sweet corn (*Zea mays* L.) as influenced by drip irrigated and planting methods. *Agric. Water Mgt., 55*(2), 85–91.

Wali, B. M., Palled, B. Y., Kalaghatagi, S. B., Babalad, H. B., & Megeri, S. N., (1988). Response of castor genotype to irrigation and nitrogen. *Maharashtra Agric. Univ. J., 16*(2), 262–263.

Watson, D. F., (1958). The dependence of net assimilation rate on leaf area index. *Ann. Bot., 10*, 41–71.

Weiss, E. A., (1971). *Castor, Sesame and Safflower.* Leonard Hill, London, p. 123.

Weiss, E. A., (1983). *Oilseed Crops.* Longman Hill, London, pp. 45–50.

Wentworth, S., & Jacobs, B., (2006). Sustainable irrigated maize on sodic soils in the Lachlan Valley. In: *Paper Presentation at Australian Agronomy Conference* (p. 7). Australian Society of Agronomy, Sydney – AU.

Willey, R. W., (1979). Intercropping, importance and research need competition and yield advantages. *Field Crop Abstract, 32*(1), 1–10.

Williams, R. F., (1946). Physiology of plant growth with special reference to the concept of net assimilation rate. *Ann. Bot., 1,* 41–72.

Wilson, R. F., Burke, J. J., & Quisenberry, J. E., (1987). Plant morphological and biochemical responses to field water deficits, Part II: Response of leaf glycerolipid composition in cotton. *Plant Physiol., 84,* 251–254.

Woldesadik, K., Gertson, U., & Ascard, J., (2003). Shallot yield quality and storability as affected by irrigation and nitrogen. *J. Hort. Sci. Biotech., 784,* 549–553.

Yadav, R. L., Sen, N. L., & Yadav, B. L., (2003). Response of onion to nitrogen and potassium fertilization under semiarid condition. *Indian J. Hort., 60,* 176–178.

Yazar, A., Sezen, S. M., & Gencel, B., (2002). Drip irrigation of corn in the Southeast Anatolia Project (GAP) area in Turkey. *Irrig. and Drain., 51,* 293–300.

Yellamandareddy, T., & Sankarareddy, G. H., (2005). *Efficient Use of Irrigation Water.* Kalyani Publishers, Ludhiana, p. 118.

Zeng. D. Q., Brown, P. H., & Holtz, B. A., (2000). Potassium fertigation improves soil K distribution, Builds Pistachio yield and quality. *Fluid J.,* 1–2.

APPENDIX I

Drip System Cost for Castor + Onion Intercropping System
(Area of one ha)

S. No.	Details of the drip system accessories	Quantity required	Unit cost (Rs.)	Total cost (Rs.)
1	Main line PVC pipe (75 mm)	3	375	1125
2	Main line PVC pipe (63 mm)	12	250	3000
3	Sub main line – PVC (50 mm)	30	200	6000
4	In line lateral (16 mm)	6666	8	53328
5	Flush valve	12	50	600
6	Ball valve (63 mm)	6	200	1200
7	Ball valve (50 mm)	6	175	1050
8	Filter 2" (75 mm)	1	2500	2500
9	Ventury assembly	1	1200	1200
10	GTO (16 mm)	135	8	1080
11	PVC fitting and accessories	1500	1	1500
12	Labor charges	2000	1	2000
	Total cost Rs. (US$)			**74583 (1243 US$)**
	Depreciation @ 15 percent	11187		
	Interest rate @ 8 percent	5966.64		
	Repair and maintenance cost	1000		
	Total, Rs. (US$)	**92737 (1545 US$)**		

APPENDIX II

Cost of Cultivation of Castor + Onion Intercrop (Rs ha^{-1}) Under Surface Irrigation.

S. No.	Details	Rs./unit	Unit rate (Rs.)	Total cost (Rs.)
	Castor			
1	Seed @ 5 kg/ha	200	5	1000.00
2	Land preparation			
	Tractor plowing, harrowing, and leveling	1750	3	5250.00
3	Manure (FYM)	500	12.5	6250.00
4	Recommended dose of fertilizer			
	i) 130 kg Urea (60 kg Nitrogen)	7	130	913.00
	ii) 187.5 kg Superphosphate (30 kg P_2O_5)	6	187.5	1125.00
	iii) 50 kg muriate of potash (30 kg K_2O)	7	49.8	348.60
5	Manure and fertilizer application	200	8	1600.00
6	Sowing, thinning operation	200	10	2000.00
7	Micro-nutrient and amendments	200	3	600.00
8	Irrigation	200	5	1000.00
9	Insecticides	550	4	2200.00
10	Insecticide application	250	4	1000.00
11	Intercultural operation	200	10	2000.00
12	Harvesting	200	12	2400.00
13	Threshing and winnowing	1000	1	1000.00
	Total cost A, Rs. (US$)			**28686.60 (478 US$)**
	Onion intercrop			
1	Seed in kg/ha	332	45	14940.00
2	Sowing	200	10	2000.00
3	Manure topdressing	200	5	1000.00
4	Micro-nutrient and amendments	200	3	600.00
5	Insecticides	550	4	2200.00
6	Insecticide application	250	4	1000.00
7	Harvesting	200	15	3000.00
8	Postharvest	200	8	1600.00
	Subtotal cost (B), Rs. (US$)			**26340.00 (439 US$)**
	Grand total (A+ B), Rs. (US$)			**55026.60 (439 US$)**

CHAPTER 16

AUTOMATED AND NON-AUTOMATED FERTIGATION SYSTEMS FOR CUCUMBER INSIDE A POLYHOUSE

ANJALY C. SUNNY and V. M. ABDUL HAKKIM

ABSTRACT

Automated fertigation system is a highly advanced system for water and fertilizer administration in irrigated agriculture. It promises the application of water in right quantity along with right fertilizer at right time, thereby reducing fertilizer loss and labor resulting in the saving of money with the help of an automated mechanism. The present study was undertaken to evaluate the performance of a timer based automated fertigation system with a FIP. Field evaluation of the developed automated fertigation system was carried out by growing salad cucumber variety 'Saniya' inside a polyhouse located at Agricultural Research Station, Anakkayam. A comparative evaluation was carried out between biometric observations and yield parameters of the two sets of crops: one fertigated automatically with the developed system the other one fertigated using venturi injector. Data collected were subjected to ANOVA (Analysis of Variance) and Student-t-test. The main crop growth parameters like the height of the plant, days to first flowering, days to 50% flowering, days to initial budding, days to first harvest and leaf area index were observed. Yield parameters viz. size of the fruit, number of fruits harvested per plant and average yield were recorded during the study. Values of all these parameters were found to be better for the crops grown inside the polyhouse with automated fertigation compared to the other.

16.1 INTRODUCTION

The adoption of fertigation by farmers largely depends on the benefits derived from it and fertigation is in its introductory stage in Kerala. Its success in terms of improved production depends upon how efficiently plants take up the nutrients. Proper scheduling and intervals are also needed to provide nutrients at a time when plants require them. The adoption of fertigation worldwide has shown favorable results in terms of fertilizer use efficiencies and quality of produce besides the environmental advantages. The choice of selecting various water-soluble fertilizers are enormous and therefore, selection of chemicals should be based on the property of avoiding corrosion, softening of the plastic pipe network, safety in field use and solubility in water.

Automated fertigation system is a highly advanced system of drip automation for water and fertilizer administration in agriculture. It promises the application of water in right quantity with right fertilizer at right time, without manual endeavors and labor. Thus, labor cost can be reduced with the help of an automated mechanism. An automated fertigation system can help producers to make correct choices that can essentially affect water and fertilizer utilization and can decrease fertilizer loss. Some automated systems are capable of integrating irrigation scheduling with nutrient dosing activities while other systems only manage the nutrient dosing equipment.

This chapter focuses on the comparative evaluation of automated and non-automated fertigation systems inside the polyhouse.

16.2 MATERIALS AND METHODS

Polyhouse for this experiment was made using GI class B pipe poles. The roofing was provided with a transparent UV (Ultra Violet) stabilized low-density polyethylene sheets of 200-micron thickness, which created a microclimate inside the polyhouse by regulating relative humidity and temperature, as it partially cuts the UV rays. The specifications of the polyhouse used for the study are shown in Table 16.1.

TABLE 16.1   Specifications of Polyhouse

Particulars	Specifications
Center height	6.5 m
Side height	4 m
Area inside	291.9 m²
GI pipes	Class B of 2-inch diameter
Roofing	200-micron thickness UV stabilized LDPE
Side net	40 mesh nylon insect proof net

16.2.1   CROP AND VARIETY

Salad cucumber (*Cucumis sativus*) variety Saniya was used for the experiment. Seeds were sown in pro tray containing a mixture of vermicompost and coir pith in 1:1 ratio to a depth of 0.5 cm. These seedlings were transplanted into grow bags on the seventh day.

16.2.2   EXPERIMENTAL PROCEDURE

Evaluation of the automated fertigation system was carried out by installing polyhouse of 291.9 m². Total 186 plants were planted in the polyhouse and were automatically fertigated; another 24 plants were planted in the same polyhouse, which was fertigated using venturi injector. The biometric and yield parameters of randomly selected plants, 4 and 7 in number respectively from each plot were noted and were compared with each other to evaluate the efficiency of the system using statistical analysis.

16.2.3   LAYOUT OF THE EXPERIMENT

The first set of plants with automated fertigation system were grown inside the polyhouse in seven rows at a spacing of 2 x 1.5 m with 24 plants in one row and 27 plants in the other six rows adding to a total number of 186 plants. The next set of 24 plants in a single row, fertigated using venturi injector was grown in the same polyhouse. All plants were grown in grow bags of size 24 x 24 x 40 cm with potting mixture, which contained soil, coir pith and dried farmyard manure (FYM) in the ratio 2:1:1. Drip irrigation system with an emitter spacing of 1.5m was installed in all plots with arrow drips of 8 lph capacity.

16.2.4   AUTOMATED FERTIGATION SYSTEM

The fertigation system was installed inside the polyhouse. The required amount of different fertilizers for the plant was filled in separate fertilizer tanks and the tank was filled with desired quantity of water with the help of push button switch. Fertilizers used were ammonium nitrate (NH_4NO_3), mono-ammonium phosphate ($NH_4H_2PO_4$) and potassium sulfate (K_2SO_4). Inside each tank, these fertilizer solutions were mixed thoroughly with the help of a bubbler. After mixing, the solutions were pumped to the mixing tank sequentially according to the preset timings from where it was pumped to the drip system through FIP. Other nutrient fertilizers such as calcium nitrate ($Ca(No_3)_2$) essential for the plant growth, were directly fed into the mixing tank in the form of solutions whenever necessary.

16.2.5   FIELD DATA COLLECTION

16.2.5.1   BIOMETRIC OBSERVATIONS

Biometric analysis on growth of the plant was done. The main crop growth parameters like height of the plant, days to initial budding, days to first flowering, days to 50% flowering, days to first harvest, Leaf Area Index (LAI) were observed. Biometric observations of 4 randomly selected plants were taken from each plot.

Plant height was measured from ground level to tip of topmost leaf. Readings were recorded for each selected plant from three different treatment plots from the transplanted date at an interval of 18 days.

The time taken by the crop to start initial budding stage from date of transplanting was observed. The number of days for each treatment was recorded.

The time taken by the crops from initial budding to start initial flowering stage from date of transplanting was observed. The number of days was recorded for each treatment.

The time by which, 50% of the plants got its flowers from date of transplanting was observed. The number of days for each treatment was recorded.

The time by which first fruit was seen from date of transplanting was observed. The number of days for each treatment was recorded.

The number of days taken by the crops to reach final fruiting stage for the first harvest was recorded for each treatment.

The average length and width of five leaves of the selected plants were taken from the date of transplanting at an interval of 18 days and the Mean Leaf Area (LAm) and in turn, the leaf area index (LAI) was found out by the method of estimation suggested by Blanco and Folegatti (2003).

$$LA = [0.859 * (L * W)] + 2.7 \qquad (1)$$

$$LAI = [(LAm * N)] / A \qquad (2)$$

where: L, W are the average of length and width of the leaves of the selected plant; N is the number of leaves in that plant, and A is the area occupied by the plant.

Yield parameters like size of the fruit, number of fruits harvested per plant and yield of seven plants were recorded during the study.

Seven plants were selected randomly from each plot. The total number of fruits per plant was recorded at each harvest and the added total number at the end of the crop was calculated as the yield of randomly selected plants. Also, the length and equatorial circumference of each fruit obtained was measured and average for each plant was calculated.

Harvesting of the crop was done in each plot after attaining maturity. Weight of harvested fruits was taken and the yield was worked out in t/ha.

16.2.6 STATISTICAL ANALYSIS

The data collected was subjected to ANOVA (Analysis of Variance) and Student-t-test and executed using the software SYSTAT and MS Excel. CRD design was used for the analysis. Wherever the results were significant, critical differences were worked out at probability level $p < 0.05$. The non-significant differences were denoted as NS. With respect to Student t-test, if the calculated value exceeds the table value, then the treatment is significantly different at that level of probability based on the hypothesis tested. In the present study, it was considered a significant difference at $p = 0.05$, and this means that if the null hypothesis were correct (i.e. the treatments do not differ) then "*t*" value has to be greater as this, on less than 5% of occasions. This means that the treatments do differ from one another, but we still have nearly a 5% chance of being wrong in reaching this conclusion.

16.3   RESULTS AND DISCUSSION

Comparative evaluation was carried out for biometric observations and yield parameters between two sets of treatments inside the polyhouse: one fertigated automatically with the developed system (T_1) and the other one fertigated using venturi injector (T_2) at various stages of plant growth. The observations were taken once in a week from both the plots.

16.3.1   BIOMETRIC OBSERVATIONS

Drip fertigation enables the application of soluble fertilizers and other chemicals along with irrigation water in the vicinity of the root zone (Narda et al., 2002; Patel et al., 2011). The application of water and nutrients in small doses at frequent intervals in the crop root zone ensures their optimum utilization and higher growth (Jayakumar et al., 2014). The results (Table 16.2) show that at the final stages, plant height was significant between the individual treatments and T_1 outperformed T_2. It registered the maximum plant height of 273 cm at the 4th observation, followed by T_2 with 242.8 cm.

TABLE 16.2   Influences of Different Treatments on Plant Height of Cucumber at Various Stages of Growth

Plant height (cm)	Observations			
	1st	2nd	3rd	4th
T_1	18.5	77.8	159.3	273.0
T_2	22.0	62.0	142.8	242.8
T_1 Vs T_2 (t value)	NS	NS	4.34**	6.58**

**Significant at $p < 0.05$; NS – Non-significant.

a. **Flowering parameters:** Earliest flowering was observed in the treatment T_1 (21 days), whereas in the treatment T_2, it was late by 3 days as shown in Table 16.3. The optimum levels of nutrient status in the media aided early flowering and the increase in number of pistillate flowers might be due to the vigorous vine growth and more number of branches resulting in increased metabolic activity in cucumber (Bishop, 1969). Similar trend was observed for 50% flowering, first fruit and first harvest in T_1 and which was followed by T_2.

TABLE 16.3 Date of Occurrence of Different Flowering Parameters

Events	T1	T2
First flower bud	27–12–15	28–12–15
First flowering	04–01–16	07–01–16
50% flowering	07–01–16	09–01–16
First fruit	06–01–16	09–01–16
First harvest	15–01–16	21–01–16

b. **Leaf Area Index:** The results (Table 16.4) indicate that at all the stages; the values of T_1 were numerically higher, compared to T_2. This indicated that fertigation may give maximum leaf growth for cucumber. The vegetative growth of the plant is directly related to the nitrogen applied (Klein et al., 1969). Moreover, according to studies conducted by Baruah and Mohan (2008), potassium application is important in leaf growth and development. Nitrogen, phosphorus, and potassium are three necessary nutrients, which affect the plant growth and thus the uniform and frequent application of fertilizer through automated drip fertigation system, which might have resulted in better leaf area index.

TABLE 16.4 Influences of Different Treatments on LAI of Cucumber Plant at Three Stages of Growth

LAI	2nd	3rd	4th
T_1	15.80	36.90	58.6
T_2	9.01	17.19	36.9
T_1 Vs T_2 (t value)	7.89**	2.53**	4.229**

**Significant at $p<0.05$; NS – Non-significant.

16.3.2   YIELD PARAMETERS

a. **Number of fruits per plant:** The results (Table 16.5) show that T_1 recorded higher number of fruits per plant than T_2 and differences were statistically significant. It registered maximum number of 29.12 fruits per plant and this was followed by T_2 with 10.50 fruits. The increase in number of fruits in T_1 might be due to the increased vegetative growth of plants under the developed automatic system leading to enhanced nutrient uptake and better water utilization, which resulted in increased rate of photosynthesis and translocation of nutrients into the reproductive part or the produce compared to the conventional method of fertilizer application. The present findings agree with the results of Sharma et al. (2011). According to Ramnivas et al. (2012), interaction between irrigation and fertigation might have resulted in maximum fruit weight.

TABLE 16.5   Influences of Different Treatments on Number of Fruits per Plant of the Cucumber

Treatments	No. of fruits/plants
T_1	29.12[a]
T_2	10.50[b]
SEd	2.266
CD (P = 0.05)	5.388

b. **Size of the fruit:** The results (Table 16.6) show that T_1 recorded the higher fruit weight than T_2. It registered the maximum fruit weight of 246.4 g and this was followed by T_2 with 212.9 g. Table 16.7 shows that T_1 registered the maximum fruit length of 21.35 cm and it was followed by T_2 with 20.70 cm. The increase in length of the fruit might be due to frequent water and nutrient supply through drip fertigation so that crop plants could complete all metabolic process at appropriate time. The adequate moisture and moisture supply also help in keeping various enzyme systems active. Therefore, quality of the produce is better in drip fertigated crops as compared to the control.

TABLE 16.6 Influences of Different Treatments on Weight of the Cucumber Fruit

Treatments	Average weight of the single fruit (g)
T_1	246.4[a]
T_2	212.9[b]
SEd	13.063
CD(P=0.05)	27.44

TABLE 16.7 Influences of Different Treatments on Length of the Cucumber Fruit

Treatments	Length (cm)
T_1	21.35[a]
T_2	20.70[a]
SEd	0.77
CD (P = 0.05)	1.62

The results (Table 16.8) show that the T_1 recorded the higher equatorial circumference than T_2. It registered the maximum equatorial circumference of 16.25 cm and this was followed by T_2 with 12.75 cm. This is because of the increase in crop growth due to the interaction effect between irrigation and fertigation levels. The 100 percentage applications of the scheduled nutrients to the root zone had also contributed to the fruit diameter (Ramnivas et al., 2012). These findings are in agreement with those by Singh and Singh (2012), who found that the trickle irrigation with 100% recommended nitrogen fertilizer gave the maximum fruit circumference, fruit length and fruit weight of papaya.

c. **Total yield:** The results (Table 16.9) show that T_1 recorded the higher fruit yield of 23.86 t ha[-1] and this was statistically significant over T_2 with 7.71 t ha[-1]. This might be due to the combined effect of cultivars, wider spacing, polyhouse cultivation and timely and uniformly availability of all the nutrients through the developed automated fertigation system. The present results are in agreement with the findings of Arora et al. (2006) in greenhouse grown tomato; and Ban et al. in melons (2006). Automated drip fertigation of cucumber adequately sustains favorable vegetative and reproductive growth as compare to conventional method of fertilizer application.

TABLE 16.8 Influences of Different Treatments on Equatorial Circumference of the Cucumber

Treatments	Equatorial circumference (cm)
T_1	16.25
T_2	12.75
CD (P=0.05)	NS

TABLE 16.9 Influences of Different Treatments on Total Yield of the Cucumber Fruit

Treatments	Total Yield (t/ha)
T_1	23.86[a]
T_2	7.71[b]
SEd	1.16
CD (P=0.05)	2.44

16.4   SUMMARY

Crop growth parameters like plant height, days to initial budding, days to 50% flowering, days to first fruit, days to first harvest and leaf area index and the yield parameters (such as number of fruits per plant, weight of the fruit, length of the fruit, equatorial circumference of the fruit) and total yield in t/ha were observed for two treatments: T_1 = crop grown inside the polyhouse and fertigated using the developed system, and T_2 = crop grown inside the polyhouse fertigated using venturi injector. The results indicate that the T_1 outperformed T_2 in case of all parameters. From the present study, it can be inferred that the automated fertigation system installed inside the polyhouse can be considered as the best treatment as it gave the maximum value of yield parameters and biometric observations. Thus it can be concluded that the developed system for automatic fertigation ensured better yield for cucumber variety 'Saniya' grown inside the polyhouse.

KEYWORDS

- automation
- coir pith
- cucumber
- farmyard manure
- fertigation
- leaf area index
- polyhouse
- venture
- vermicompost

REFERENCES

Arora, S. K., Bhatia, A. K., Singh, V. P., & Yadav, S. P. S., (2006). Performance of indeterminate tomato hybrids under greenhouse conditions of north Indian plains. *Haryana J. Hortic Sci., 35*(3 & 4), 292–294.

Ban, D., Goreta, S., & Borosic, J., (2006). Plant spacing, and cultivar affect melon growth and yield components. *Sci. Hortic., 109*, 238–243.

Baruah, P. J., & Mohan, N. K., (2008). Effect of potassium on LAI, phyllochrome, and number of leaves of banana. *Banana News Letter, 14*, 21–22.

Bishop, R. F., Chipmon, E. W., & Mae-eachern, C. R., (1969). Effect of nitrogen, phosphorous, and potassium on yield and nutrient levels in laminate and petioles of pickling cucumber. *Can. J. Soil Sci., 49*, 297–404.

Blanco, F. F., & Folegatti, M. V., (2003). A new method for estimating the leaf area index of cucumber and tomato plants. *Hortic. Brasileira, 21*(4), 666–669.

Jayakumar, M., Surendran, U., & Manickasundaram, P., (2014). Drip fertigation effects on yield, nutrient uptake and soil fertility of Cotton in semi-arid tropics. *Int. J. Plant Prod., 8*(3), 375–389.

Klein, L., Levin, L., Bar-Yosef, B., Assaf, R., & Berkovitz, A., (1989). Drip nitrogen fertigation of 'Starking Delicious' apple trees. *Plant and Soil, 119*(2), 305–314.

Narda, N. K., & Chawla, J. K., (2002). A simple nitrate sub-model for trickle fertigated potatoes. *Irrig. Drain., 51*, 361–371.

Patel, N., & Rajput, T. B. S., (2011). Simulation and modeling of water movement in potato (*Solanum tuberosum*). *The Indian J. Agric. Sci., 81*, 25–32.

Ramnivas, K., Sarolia, R. A., Pareek, S., & Singh, V., (2012). Effect of irrigation and fertigation scheduling on growth and yield of guava (*Psidiumguajava* L.) under meadow orcharding. *Afr. J. Agric. Res., 7*(47), 6350–6356.

Sharma, A., Kaushik, R. A., & Sharma, R. P., (2011). Response of cultivars, plant geometry and methods of fertilizer application on parthenocarpic cucumber (*Cucumissativus* L.) under zero energy polyhouse condition. *Veg. Sci.*, *38*(2), 215–217.

Singh, H. K., & Singh, A. K. P., (2005). Effect of refrigeration on fruit growth and yield of papaya with drip irrigation. *Environ. Ecol.*, *23*, 692–695.

CHAPTER 17

PERFORMANCE OF DRIP IRRIGATED CAULIFLOWER USING A POLYLACTIC ACID ROLL PLANTER

K. VAIYAPURI, B. J. PANDIAN, and SELVARAJ SELVAKUMAR

ABSTRACT

The present investigation evaluated the effects of PLA roll planter on growth, yield, and water productivity of cauliflower under drip irrigation. Drip fertigation with roll planter gave significantly increased plant height during 20 days after planting to harvest stage (50.70 cm) due to moderate drainage and better aeration because of knitted fabrics. Therefore, the balance between water and air for the root zone was well maintained. It promoted good, healthy growth of plants. It was followed by roll planter with hand irrigation (50.40 cm). Farmer's method of cultivation recorded the lowest plant height at all stages. This treatment gave 56% more leaf area index compared to farmers' method of cultivation. Drip fertigation with roll planter recorded 60.29% increase in yield compared farmers' method due to an optimum number of leaves and more photosynthetic activities that enhanced dry matter accumulation and better mobilization of plant nutrients during later stages of plant growth thus resulting in increased cauliflower yield attributes. Drip fertigation with roll planter and drip fertigation with native soil and roll planter with hand irrigation showed less water consumption (300 to 340 mm) and higher water productivity (0.04 kg/m³ in 80 m² area) compared with farmer's method (500mm & 0.01 kg/m³).

17.1 INTRODUCTION

Efficient utilization of available water resources is crucial for India, which shares 17% of the global population with only 2.4% of land and 4% of the water resources. The per capita water availability, in terms of average utilizable water resources in the country, is presently 1750 m³ that is expected to dwindle down to 760 m³ in 2050. All these factors emphasize the need for water conservation and improvement in water use efficiency. Adoption of micro irrigation is one of the answers to this problem. Drip Irrigation is one of the Water Saving Technology with 95% efficiency, besides it increases the productivity of crops. Lack of awareness, initial high cost, and poor technological backup are some of the reasons for non-adoption. Many alternative products are being used to suit small-scale farmers to achieve the concept of increasing water use efficiency. One such initiative was taken by a Japan Company Mitsukawa & Co. Ltd developed PLA Roll Planter.

India is the second largest producer of vegetable crops in the world. However, the production is far below the requirement of a balanced diet to every individual. To cater the future vegetable needs in India, the present production of 156.33 million tons must be raised to 225 million tons by 2020 and 350 million tons by 2030 (Anonymous, 2011). Bringing additional area under vegetables, using hybrid seeds and the use of improved agro-techniques are different ways to achieve this target. Another potential approach is the promotion of protected cultivation of vegetable crops (Rajaseker et al., 2013).

PLA (Poly Lactic Acid) roll planter is an agricultural material knitted in a tubular shape and is used in the form of packing the soil and sand in the cylindrical knit. It has a moderate drainage and breathability because it is made of knitted fabrics. Therefore, the balance of the water and air for the root zone in roll planter is well maintained, which promotes good and healthy growth of plants. The materials are biodegradable fibers. The main benefits include effective irrigation, good air, water balance, no soil deification and prevent root temperature rise by evaporation heat.

The present investigation evaluated the effects of PLA roll planter on growth, yield, and water productivity of cauliflower under drip irrigation.

17.2 MATERIALS AND METHODS

Field experiment was conducted in farmer's field at Semmedu near ISHA Meditation Centre in Coimbatore during winter season of January 2016 to March 2016 with an objective to increase the cauliflower yield and water productivity under PLA roll planter technique. The treatments were:

- T_1 – PLA Roll planter with drip fertigation;
- T_2 – Conventional drip fertigation;
- T_3 – PLA Roll planter with hand irrigation; and
- T_4 – Native land with hand irrigation.

PLA (Poly Lactic Acid) roll planter is an agricultural material knitted in a tubular shape and is used in the form of packing the soil and sand in the cylindrical knit. The experimental plot size was 80 m^{-2}. The experiment site was black soil with clay loam with a pH of 7.5, bulk density of 1.27 g cm^{-3} and electrical conductivity of 0.28 dSm^{-1}, respectively. The soil depth was 90 cm with infiltration rate and organic carbon of 0.7 cm/h and 0.58%, respectively. The soil texture was clay with 15.75% coarse sand, 38.75% silt and 45% clay with medium depth. The moisture content at field capacity, permanent wilting point, and available soil moisture were 41.28, 20.27 and 21.01%, respectively. The test crop was cauliflower (Figure 17.1). The seedlings were transplanted as per the treatments. Fertigation and fertilizer were applied per treatment schedule (Table 17.1).

The plant protection chemicals were sprayed using recommendations in the TNAU crop production guide 2015 (TNAU, 2015). The plant height at periodical intervals, number of leaves, leaf length, leaf width, root length, root volume, root weight, and shoot length were recorded. Yield attributes (viz.: number of fingers per head, curd length, curd width, and curd weight were recorded. Water productivity was calculated in kg ha^{-3}.

17.3 RESULTS AND DISCUSSION

17.3.1 GROWTH CHARACTERISTICS

The growth parameters are presented in Table 17.2. The treatment drip fertigation with roll planter had recorded significantly increased plant height during 20 days after planting to the harvest stage (50.70 cm) due to moderate drainage and better aeration because of knitted fabrics. Therefore, the balance of the water and air for the root zone was well maintained. The plants were found to be good with healthy growth of plants. It was followed by roll planter with hand irrigation (50.40 cm). Farmer's method of cultivation recorded the lowest plant height at all the stages.

A. Before transplanting of cauliflower

B. After transplanting and imposing the treatment

FIGURE 17.1   (See color insert.) Cabbage vegetable crop in the field.

TABLE 17.1 Fertigation Schedule for 80 m² Area: Details of Treatments

Treatment	Crop stage	Duration	Fertilizer applied	Quantity
T₁: Roll planter with drip fertigation (80 m² area)	Transplanting to plant establishment	10 days	ALL 19	500 g
			Potassium Nitrate	60 g
			Urea	122 g
	Curd Initiation Stage	30 days	Potassium Nitrate	890 g
			Mono Ammonium Phosphate	1000 g
	Curd Development	45 days	Urea	1200 g
			Sulfate of Potash	965 g
T₂: Native soil with drip fertigation (80 m² area)	Transplanting to plant establishment	10 days	ALL 19	500 g
			Potassium Nitrate	60 g
			Urea	122 g
	Curd Initiation Stage	30 days	Potassium Nitrate	890 g
			Mono Ammonium Phosphate	1000 g
	Curd Development	45 days	Urea	1200 g
			Sulfate of Potash	965 g

Fertilizer application doses

Treatment	Time of application	N	P	P	Others
			Fertilizer dose		
T₃: Roll planter with native soil (80 m² area)	During planting (as basal)	400 g	800 g	400 g	NIL
	45 days after planting (top dressing)	400 g	NIL	NIL	16 g of Micronutrient mixture
T₄: Farmers method (80 m² area)	During planting (as basal)	400 g	800 g	400 g	NIL
	45 days after planting (top dressing)	400 g	NIL	NIL	16 g of Micronutrient mixture

*Fertilizers were applied with a minimum distance of 15cm from the plant.

Leaf area index was recorded at 10 DAP up to the harvest stage (Table 17.2). Drip fertigation with roll planter gave significantly higher leaf area index for all stages (0.50, 2.97, 13.12 and 19.95). Farmers' method of cultivation recorded the lowest leaf area Index for all stages. The leaf area index was 56% higher compared with farmers' method of cultivation. The results were in accordance with the findings of Babul et al., 1998] in cauliflower. Foliage weight (without curd), shoot length, root length, root volume, and total biomass were recorded during the harvest stage.

The treatment drip fertigation with roll planter had recorded significantly lower foliage weight compared to other treatments (510 g) due to conversion of economic parts. Drip fertigation with native soil and roll planter with hand irrigation recorded the higher foliage weight due to continuous moisture availability and vigorous vegetative growth. Similar trend was observed for shoot length.

TABLE 17.2   Effect of Treatments on Growth Characteristics of Cauliflower Under Drip Irrigation

Treatment	Plant height (cm) at DAP				Leaf Area Index at DAP			
	10	20	40	Harvest	10	20	40	Harvest
T₁ – Drip fertigation with roll planter	11.00	19.80	30.05	56.70	0.50	2.97	13.12	19.95
T₂ – Drip fertigation with native soil	8.65	16.10	25.10	45.20	0.41	2.36	7.79	10.65
T₃ – Roll planter with hand irrigation	12.45	17.10	29.35	50.40	0.68	2.26	11.89	16.33
T₄ – Farmer's method	8.25	12.60	22.40	38.60	0.35	1.01	6.89	8.65
SEd	0.53	1.06	1.43	2.09	0.01	0.08	0.31	0.48
CD (p = 0.05)	1.14	2.27	3.07	4.49	0.03	0.17	0.66	1.03

17.3.2   ROOT CHARACTERISTICS

About root length, root weight and root volume were recorded in respective treatments (Table 17.3). Farmers' method of cultivation recorded significantly higher root length, root weight and root volume (20.40, 76.00 cm and 68.00 cc) and it was followed by drip fertigation with native soil due to more anchorage and proliferation to deeper soil for nutrient and moisture. Drip fertigation with roll planter recorded lower root length, root weight and root volume due to roots penetrated and anchorage only in the rooting medium of roll planter and more aeration, more drainage, and better aeration.

17.3.3   YIELD ATTRIBUTES AND YIELD

The yield attributes (*viz.*, curd weight; curd width, curd length and number of fingers per head) recorded significantly higher values in drip fertigation with roll planter (872, 33.70, 13.40 cm and 29.20) as shown in Table 17.4 and Figure 17.2. It was followed by roll planter with hand irrigation due to conversion of vegetative parts to economic parts and the balance of the water and air for the root zone in roll planter. The main factors include effective irrigation, good air, water balance, no soil deification, and root temperature rise by evaporation heat. Roll planter with hand irrigation was second best treatment. Farmers' method recorded the lower yield attributes (363, 21.60, 10.20 g). Similar trend was recorded in yield. The treatment drip fertigation with roll planter recorded 60.29% yield increase over farmers' method due to optimum number of leaves and more photosynthetic activities that enhanced food accumulation and better mobilization of plant nutrients during later stages of plant growth thus resulting in increased the cauliflower yield. These findings agree with the finding of in potato by Vivek et al. (2011).

TABLE 17.3   Effect of Treatments on Root, Shoot Length, Root Volume, and Total Plant Weight of Cauliflower under Drip Irrigation

Treatment	Foliage weight (g)	Shoot length (cm)	Root length (cm)	Root weight (g)	Root volume (cc)	Single plant biomass weight (Kg)
T_1 – Drip fertigation with roll planter	509.96	46.00	16.20	27.00	25.00	1.39
T_2 – Drip fertigation with native soil	682.95	58.60	17.60	51.00	44.00	1.15
T_3 – Roll planter with hand irrigation	680.95	50.40	15.80	33.00	29.00	1.22
T_3 – Farmer's method	641.96	38.20	20.40	76.00	68.00	1.05
SEd	16.42	1.33	0.48	1.89	1.67	0.03
CD (p = 0.05)	35.22	2.85	1.02	4.05	3.58	0.07

17.3.4   WATER PRODUCTIVITY

Total water application and water productivity were determined. Drip fertigation with roll planter and drip fertigation with native soil and roll planter with hand irrigation showed less water consumption (300 to 340 mm) and more water productivity (0.04 kg m^{-3} in 80 m^2 area) compared to farmer's method (500 mm and 0.01 kg m^{-3} in 80 m^2 area).

TABLES 17.4   Effect of Treatments on Yield Attributes and Water Productivity in 80m² of area of Cauliflower Under Drip Irrigation

Treatment	Curd weight (gm)	Curd width (cm)	Curd length (cm)	Fingers/ Head (No)	Yield (kg/80 m²)	Total water applied (mm)	Water productivity (kg/m³)
	per head				in 80 m² of area		
T₁ – Drip fertigation with roll planter	871.94	33.70	13.40	29.20	136	340	0.04
T₂ – Drip fertigation with native soil	427.97	25.90	14.20	23.40	62	340	0.02
T₃ – Roll planter with hand irrigation	515.96	29.70	14.80	24.00	78	300	0.03
T₃ – Farmer's method	362.98	21.60	10.20	21.20	54	500	0.01
SEd	20.26	0.77	0.36	0.65	3.20		
CD (p = 0.05)	43.47	1.64	0.77	1.39	6.87		

17.4   SUMMARY

Increased plant height at harvest stage under roll planter with drip fertigation (56.70 cm) was recorded compared to farmer's method of cultivation (38.60 cm). It was followed by roll planter with hand irrigation (50.40 cm). The plant weight shoot length, root length, root weight, and volume values were higher under roll planter with drip fertigation system compared to farmer's method. The curd weight, curd width, curd length, number of fingers values were higher under roll planter under drip fertigation over farmer's method and it was followed by roll planter with hand irrigation. It was further observed that the water consumption per plot was less (340 mm) and water productivity per plot was more (0.04 kg m⁻³) under roll planter with drip fertigation compared to farmer's method (500 mm and 0.01 kg m⁻³ of water consumption and water productivity, respectively).

ACKNOWLEDGMENT

Mitsukawa & Co. Ltd, Japan, and Japan International Collaborative Agency for funding the project is gratefully acknowledged.

Field observation

Discussion with JICA

FIGURE 17.2 **(See color insert.)** Field harvesting of cabbage.

KEYWORDS

- cauliflower
- water productivity
- yield attributes

REFERENCES

Anonymous, (2011). Vision 2030. Indian Institute of Vegetable Research, Varanasi, U.P., p. 10.

Babul, A. S., Verma, R. M., Panchbhai, D. M., Mahorkar, V. K., & Khankhane, R. N., (1998). Effect of biofertilizer and nitrogen levels on growth and yield of cauliflower. *Orrisa Journal of Horticulture, 26*(2), 14–17.

Crop Production Guide, (2015). Tamil Nadu Agricultural University, Coimbatore–3, online.

Rajasekar, M., Arumugam, T., & Ramesh, K. S., (2013). Influence of weather and growing environment on vegetable growth and yield. *Journal of Horticulture and Forestry, 5,* 160–167.

Vivek, K., Jaisawal, R. C. S., Singh, A. P., Kumar, V., Khurana, S. M. P., & Pandy, S. K., (2011). Effect of bio-fertilizer on growth and yield of potato. In: National symposium on sustainability of potato revolution in India, *Journal of the Indian Potato Association, 38*(1), 60–61.

CHAPTER 18

PERFORMANCE OF DRIP FERTIGATED BANANA UNDER POLYETHYLENE MULCHING

K. B. SUJATHA, J. AUXCILIA, K. SOORIANATHASUNDARAM,
P. MUTHULAKSHMI, PRAKASH PATIL, and R. M. VIJAYAKUMAR

ABSTRACT

The study in this chapter evaluated the effects of polyethylene (PE) mulching on the yield and quality of drip fertigated banana cv. Grand Naine. The yield (97.5 t/ha) was significantly higher in T2 (Drip irrigation + Fertigation + Micronutrient foliar spray + Bunch spray) plots with significantly higher bunch weight (32.1 kg/plant), hands per bunches (10.5) and more number of fingers per bunch (201) as compared to other treatments. Moreover, the quality of the fruit in terms of TSS (19.9) and acidity (0.26%) was the best in T2 with a greater shelf life. A peculiar characteristic of earliness in crop growth was observed in T1 (irrigation + Fertigation + Micronutrient foliar spray + Bunch spray+ mulching), where mulching was integrated with drip fertigation that reduced the days to shooting (215.5) and harvest (316.9). On the other hand, T5 (control-flood irrigation) with the conventional method of cultivation practices showed late maturity (365.4 days) or longer crop duration. The least performance with respect to growth, yield, and quality of banana was also observed in the control plots. Hence, T2 with all inputs except mulching can be considered as the best treatment in the present investigation.

18.1 INTRODUCTION

Irrigated agriculture is a productive sector that presents a high demand for water. Banana is a herbaceous perennial consuming a higher quantity of water and fertilizers (Ghavami, 1974; Robinson and Alberts, 1986). The present water demand (Rosegrant et al., 2002) coupled with soil pollution with high usage of nutrients can be effectively managed by adopting the drip irrigation and fertigation technologies that have revolutionized commercial cultivation of banana in recent years. This technology not only can increase water use efficiency (Hegde and Srinivas, 1991) but also reduces the amount of fertilizers used under fertigation (Srinivas et al., 2001).

Water use efficiency involves the amount of water used by the plants supplied through drip irrigation, the soil evaporation in terms of pan evaporation and crop factor (Ricardo Goenaga et al., 1995) and hence we must contemplate technologies that can overcome the loss of evaporation. Natural soil mulches like leaves, straw, compost etc. have been used, which can increase the soil water holding capacity and reduced the evaporation losses. However, today, plastic or polyethylene mulching have been more effective in preventing direct evaporation of moisture from the soil as they are impermeable to water, thereby limiting the water losses and soil erosion than the natural mulches over the surface. Polyethylene mulching not only reduces soil evaporation effectively but also prevents the emergence of weeds. In addition, it prevents the soil compaction, thereby, improving the soil aeration for the growth of the plant. This reveals that the practice of drip fertigation with plastic mulching has a scope of increasing the input use efficiency (Paul et al., 2008). It has been reported that mulching has also increased the banana yield and returns (Paul et al., 2008; Agarwal, 2005).

Based on these observations, the study in this chapter evaluated the effects of polyethylene mulching on the yield and quality of drip fertigated banana cv. Grand Naine.

18.2   MATERIALS AND METHODS

The experimental material was tissue cultured banana cv. Grand Naine of the uniform stage with four replications per treatment. About fifteen plants were taken per replication and the randomized block design was used with spacing of 1.8 m x 1.8 m. The different treatment components were:

   a.   Drip irrigation (80% ER at all stages)
   b.   Fertigation (80% ER)
   c.   Micronutrient foliar spray (Banana Shakti – 2% spray at 4, 5 and 6 month after planting)

d. Bunch spray of 2% SOP (I spray – after male bud removal and II spray – at 30 days after I spray)

e. Polyethylene mulching (100-micron UV stabilized black polyethylene)

Based on these components, the following different treatment combinations were used:

T_1 a + b + c + d + e
T_2 a + b + c + d
T_3 a + b + c
T_4 a + b + d
T_5 Control (Soil application of region-specific RDF (110N:35P:330K per plant) + flood irrigation)

The plant growth, yield, and quality parameters of banana were recorded to study the integrated effect of micro irrigation and polythene mulching on performance of banana cv. Grand Naine.

The trial was initiated by planting tissue culture of banana cv. Grand Naine and different treatment combinations were imposed. At the time of initiation of the trial, soil samples were collected and analyzed for their physical and chemical properties and the result of the soil analysis is given in Table 18.1. Based on the available soil NPK content from the initial soil analysis, the fertilizer dose was calculated and applied as indicated in the ready – reckoner for the targeted yield of Grand Naine banana. Soil samples were also analyzed after the harvest of the crop for each treatment (Table 18.4).

The weather data (Maximum & Minimum temperature, Rainfall, Evaporation, Sunshine hours and Relative humidity) during the crop growing period were recorded daily to calculate the daily water requirement of the crop for 80% Evaporation Replenishment (ER) (Bhattacharyya and Rao, 1985). The water requirement was calculated as follows:

$$\text{Water requirement} = [(CPE \times Kp \times Kc \times Area \times Wp)] - RF \qquad (1)$$

where: CPE = cumulative Pan Evaporation; RF = effective rainfall (mm); Kp = pan coefficient (0.75 – 0.8); Kc = crop coefficient (0.75 = initial, 1.10 = grand growth, 1.00 = latter growth); Area = spacing of the crop; Wp = wetting percentage (0.4 = wider spacing crop; 0.8 = closer spacing crop).

TABLE 18.1   Soil Nutrient Status Before the Initiation of the Experiment

Parameter	Unit	Value	Comments
Organic Carbon	%	0.41	Low
pH	—	8.06	Slightly alkaline
EC	dSm^{-1}	0.16	Non-saline
Available N	$Kg\ ha^{-1}$	199	Low
Available P(Olsen's)	$Kg\ ha^{-1}$	11	Medium
Available K	$Kg\ ha^{-1}$	400	High
Available Zn	ppm	0.78	Deficient
Available Cu	ppm	8.61	Sufficient
Available Fe	ppm	0.29	Deficient
Available Mn	ppm	4.54	Sufficient

18.3   RESULTS AND DISCUSSION

The banana yield (97.5 t/ha) was significantly higher in T2 (Drip irrigation + Fertigation + Micronutrient foliar spray + Bunch spray) with significantly higher bunch weight (31.6 kg/plant), hands per bunches (10.5) and more number of fingers (201) and high benefit-cost ratio (BCR) (3.6) (Table 18.2) compared to other treatments. Obviously, the finger weight was less in T2 (Drip irrigation + Fertigation + Micronutrient foliar spray + Bunch spray) compared to T1 (Drip irrigation + Fertigation + Micronutrient foliar spray + Bunch spray + Polyethylene mulching), T3 (Drip irrigation + Fertigation + Micronutrient foliar spray) and T4 (Drip irrigation + Fertigation + Bunch spray) because of more number of fingers in T2. Unless light is limiting (Cockshull et al. 1992), yield is mainly restricted by the number or the size of the fruit (i.e., the sink strength) rather than the supply of assimilate (i.e., the source strength). Fruit size is determined by both cell number and cell size (Bohner and Bangerth, 1988; Ho, 1992).

The rate of fruit expansion is affected by assimilate supply (Ehret and Ho, 1986), temperature (Pearce et al., 1993b) and water relations (Ho et al., 1987) that were observed in tomato by Ho (1996). The sink strength for assimilate of a tomato fruit measured by the rate of assimilate import may be related to the routes of sugar transport into the sink cells during fruit development. Enzymatic regulation of the hydrolysis of sucrose by sucrose synthase and the accumulation of starch by ADPG pyrophosphorylase may determine the rate of assimilate import in the young fruit.

TABLE 18.2 Effect of Precision Farming Practices on Yield, Quality, and BCR for Banana Cv. Grand Naine

Treatments	Yield (t/ha)	Bunch weight (kg/ plant)	Hands/ bunch	Fingers/ bunch	Finger Weight (g)	Shelf life (Days)	TSS (°B)	Acidity (%)	BCR
T1	90.7	29.4	10.7	182.6	190.9	2.0	19.5	0.33	2.5
T2	97.5	32.1	10.5	201.7	161.2	5.0	19.9	0.26	3.6
T3	88.5	28.7	10.1	174.1	193.4	5.0	20.0	0.46	2.4
T4	91.0	29.5	9.2	201.3	190.4	.02	19.5	0.40	2.9
T5	75.9	24.6	8.5	145.1	154.0	2.0	17.5	0.26	1.4
Mean	87.7	29.0	9.79	180.9	177.9	3.2	19.3	0.34	
SEd	3.60	0.79	0.17	2.36	9.16	0.05	0.35	0.01	
CD (P=0.05)	6.11**	1.82**	0.38**	8.08**	21.11**	0.11**	0.80**	0.02**	
CV, %	4.55	3.33	2.07	2.36	6.30	1.75	2.21	3.22	

Vacuolar invertase activity may determine the sugar composition of a mature fruit but may not affect the overall dry matter accumulation of a tomato fruit. These clearly explain that the yield was determined by the balance between source and sink strengths of the plant, and the quality of the fruit was determined by the transport and metabolism of sugars within the fruit (Robinson et al., 1988).

These findings are in confirmation with the present study, where there was a balanced distribution of assimilates from the source to the sink in T2 (Drip irrigation + Fertigation + Micronutrient foliar spray + Bunch spray) recording higher yield with more number of fingers and lower finger weight. The quality of the fruit was also good compared to other treatments in terms of total soluble solids (TSS) (19.9°B) and was at par with T3 (20°B) with low acidity of 0.26%that might be due to the balanced invertase activity, which determined the composition of a mature fruit.

Growth and physiology of any plant are major factors contributing to the yield and quality of the plant. In the present investigation, it was observed that higher photosynthetic activity (Table 18.3) in T2 compared to other treatments resulted in more number of fingers and hence higher yield (Palmer, 1992). The shelf-life of the fruits was also significantly high in T2 compared to other treatments. A different trend was observed in case of T1, where the parameters like phyllochron, days to shooting, days to harvest and hence crop duration were early. This earliness or fast growth in T1 should be conferred to the polythene mulching, which was absent in other treatments. However, T5, where conventional methods were followed, showed a contrasting behavior towards these parameters. The earliness in growth for polyethylene mulching treatment was due to the elevated soil temperature due to the absorption of radiations by the black plastic mulch (Streck et al., 1995). Moreover, the soil compaction is lower under black mulch, which improved the soil aeration and hence the crop growth (Streck et al., 1995). Though earliness and faster growth was observed in T1, the yield was lower than that of T2 and this difference in yield should be due to the dry matter partitioning to the fruits that would have been influenced by the elevated soil temperature in black mulches (T1). Hence, T2 (Drip irrigation + Fertigation + Micronutrient foliar spray + Bunch spray) can be considered as the best treatment with significantly higher yield contributed by more number of fingers per hand, higher photosynthetic rate, better fruit quality, and shelf-life (Tables 18.1 and 18.4).

TABLE 18.3 Effect of Precision Farming Practices on Growth and Photosynthetic Rate in Banana Cv. Grand Naine

Treatments	Photosynthetic rate (µmol CO_2/m²/s)	Leaf temperature (°C)	Phyllochron (Days)	Days to shooting (Days)	Days to harvest (Days)
T1	21.3	33.7	9.0	215.5	316.9
T2	33.4	29.1	10.3	225.7	329.0
T3	19.6	32.5	10.3	225.4	330.7
T4	22.5	32.2	10.3	230.7	325.9
T5	15.3	32.1	10.6	235.6	365.4
Mean	18.8	32.1	10.1	226.6	333.6
SEd	0.33	0.50	0.24	3.50	8.49
CD (P=0.05)	076**	1.15**	0.79**	8.20**	9.59**
CV %	2.14	0.01	2.72	2.37	3.12

TABLE 18.4 Soil Nutrient Status at the End of the Experiment

Parameter	T_1	T_2	T_3	T_4	T_5
Organic Carbon (%)	0.49 Low	0.57 Medium	0.68 Medium	0.57 Medium	0.60 Medium
pH	8.81 Alkaline	8.85 Alkaline	8.78 Alkaline	8.97 Alkaline	8.66 Alkaline
EC (dSm⁻¹)	0.40 Non-saline	0.72 Non-saline	0.90 Non-saline	0.58 Non-saline	0.72 Non-saline
Available N (Kg ha⁻¹)	291 Medium	109 Low	218 Low	202 Low	179 Low
Available P(Olsen's) (Kg ha⁻¹)	19 Medium	21.6 Medium	22.1 Medium	20 Medium	25.1 High
Available K (Kg ha⁻¹)	605 High	749 High	710 High	698 High	651 High
Available Zn (ppm)	1.77 Sufficient	1.86 Sufficient	1.21 Sufficient	1 Deficient	1.95 Sufficient
Available Cu (ppm)	3.08 Sufficient	3.20 Sufficient	3.57 Sufficient	3.84 Sufficient	3.11 Sufficient
Available Fe (ppm)	5.53 Deficient	5.21 Deficient	4.45 Deficient	4.75 Deficient	4.78 Deficient
Available Mn (ppm)	3.11 Sufficient	3.09 Sufficient	3.78 Sufficient	3.62 Sufficient	3.41 Sufficient

18.4   SUMMARY

Micro irrigation along with polythene mulching in banana have potential to improve the input use efficiency. Hence, this study was taken up at Department of Fruit Crops, HC & RI, TNAU, Coimbatore, by planting tissue culture banana cv. Grand Naine and the different treatment combinations comprising of Drip irrigation (80% ER at all stages), Fertigation, Micronutrient foliar spray (Banana Shakthi – 2% spray at 3,4 and 5 MAP), Mulching with 100 micron UV stabilized black polyethylene and Bunch spray of 2% SOP (First spray after male bud removal and second spray at 30 days after first spray) were imposed. At the time of initiation and end of the trial, soil samples were collected and analyzed for their physical and chemical properties.

The fruit yield (97.5 t/ha) was significantly higher in T2 (Drip irrigation + Fertigation + Micronutrient foliar spray + Bunch spray) with significantly higher bunch weight (31.6 kg/plant) and number of hands per bunches (10.5) and more number of fingers per bunch (201) as compared to other treatments. Moreover, the quality of the fruit in terms of TSS (19.9) and acidity (0.26%) was the best in T2 with a greater shelf – life. Although the treatments, T1, T3 and T4 were on par with T2 for TSS, yet the acidity of these fruits were higher with reduced shelf life. The photosynthetic rate and leaf nutrient content was high in T2 compared to other treatments that might have contributed to the best yield, growth, and quality of Grand Naine. A peculiar characteristic of earliness in crop growth was observed in T1 (irrigation + Fertigation + Micronutrient foliar spray + Bunch spray+ mulching), where mulching was integrated with drip fertigation reducing the days to shooting (215.5) and harvest (316.9). On the other hand, T5 (control- flood irrigation) with conventional method of cultivation practices showed late maturity (365.4 days) or lengthy crop duration. The least performance with respect to growth, yield, and quality of banana was also observed in control. Hence, T2 with all inputs except mulching can be considered as the best treatment in the present investigation.

ACKNOWLEDGMENT

The study was supported by All India Coordinated Research Project on Fruits, Indian Council of Agricultural Research, New Delhi, India.

KEYWORDS

- banana
- fertigation
- micro irrigation
- micronutrients
- polyethylene mulching

REFERENCES

Bhattacharyya, R. K., & Rao, V. N. M., (1985). Water requirement, crop coefficient and water use efficiency of "Robusta" banana under different soil covers and soil moisture regimes. *Scientia Horticulturae*, *25*(3), 263–269.

Bohner, J., & Bangerth, F., (1988). Cell number, cell size and hormone level in semi-isogenic mutants of *Lycopersicon pimipinellifolium* differing in fruit size. *Physiologia Plantarum*, *72*, 316–320.

Cockshull, K. E., Graves, C. J., & Cave, C. R. J., (1992). The influence of shading on yield of glasshouse tomatoes. *J. of Horticultural Science*, *67*, 361–367.

Ehret, D. L., & Ho, L. C., (1986). The effects of salinity on dry matter partitioning and fruit growth in tomatoes grown in nutrient film culture. *Journal of Horticultural Science*, *61*, 361–367.

Ghavami, N., (1974). Irrigation of valley bananas in Honduras, *Trop. Agri.* (Trinidad), *51*, 443–446.

Hegde, D. M., & Srinivas, K., (1991). Growth, yield, nutrient uptake and water use of banana crops under drip and basin irrigation with N and K fertilization. *Trop. Agri.*, *68*, 331–334.

Ho, L. C., (1992). Fruit growth and sink strength. In: Marshall, C., & Grace, J., (eds.), *Fruit and Seed Production, Aspects of Development, Environmental Physiology and Ecology* (Vol. 47, pp. 101–124). SEB Seminar Series, Cambridge.

Ho, L. C., (1996). The mechanism of assimilates partitioning and carbohydrate compartmentalization in fruit in relation to the quality and yield of tomato. *J. of Expt. Bot.*, *47*, 1239–1243.

Ho, L. C., Grange, R. I., & Picken, A. J., (1987). An analysis of the accumulation of water and dry matter in tomato fruit. *Plant, Cell, and Environment*, *10*, 157–162.

Palmer, J. W., (1992). Effects of varying crop load on photosynthesis, dry matter production and partitioning of Crispin/M.27 apple trees. *Tree Physiol.*, *11*(1), 19–33.

Paul, J. C., Mishra, J. N., & Pradhan, P. L., (2008). Response of banana to drip irrigation and mulching in coastalOrissa. *J. of Agricultural Engineering*, *45*(4), 44–49.

Ricardo, G., & Heber, I., (1995). Drip irrigation recommendations for plantain and banana grown on the semiarid southern coast of Puerto Rico. *J. Agric. Univ. P. R.*, *79*, 14–27.

Robinson, J. C., & Alberts, A. J., (1986). Growth and yield response of banana (cultivar 'Williams') to drip irrigation under drought and normal rainfall conditions in the sub-tropics. *Sci. Hort., 3,* 187–202.

Robinson, N. L., Hewitt, J. D., & Bennett, A. B., (1988). Sink metabolism in tomato fruit, I: Developmental changes in carbohydrate metabolizing enzymes. *Plant Physiology, 87,* 727–730.

Rosegrant, W. M., Ximing, C., & Sarah, A. C., (2002). *World Water and Food to 2020: Dealing With Scarcity.* International Food Policy Research Institute, Washington, D. C., USA, and International Water Management Institute, Colombo, Sri Lanka, p. 125.

Srinivas, K., Reddy, B. M. C., Chandrakumar, S. S., Thimmegowda, H. B., & Padma, P., (2001). Growth, yield, and nutrient uptake of Robusta banana in relation to N and K fertigation. *Indian J. Hort., 58,* 287–293

Streck, N. A., Schneider, F. M., Buriol, G. A., & Heldwein, A. B., (1995). Effect of polyethylene mulches on soil temperature and tomato yield in plastic greenhouse. *Sci. Agric., 52*(3), 587–593.

CHAPTER 19

PERFORMANCE OF TURMERIC (*CURCUMA LONGA* L.) UNDER DRIP FERTIGATION

K. S. SANGEETHA and J. SURESH

ABSTRACT

This study was conducted at College Orchard, Department of Spices and Plantation Crops, Horticultural College and Research Institute, Tamil Nadu Agricultural University, Coimbatore from 2014 to 2015 to study the effects of fertigation of N and K on growth and yield of turmeric transplants. The experiment consisted of nine treatments replicated three times in a Randomized Block Design. The results showed that the fertigation treatment with 125% levels of N and K (water-soluble fertilizers) recorded significantly superior growth parameters *viz.,* plant height (140.05 cm), number of leaves (14.83), number of tillers (9.11), leaf length (31.35 cm), leaf breadth (10.47 cm), total dry matter production (31.50 t ha^{-1}) at 210 days after planting and yield characters *viz.,* yield per plant (401.00 g), yield per plot (109.09 kg/25m^2), estimated yield (43.64 t ha^{-1}) and estimated cured rhizome yield (7.92 t ha^{-1}) of turmeric transplants. This was closely followed by fertigation with 100% levels of N and K (water-soluble fertilizers) and recorded the highest benefit-cost ratio of 3.65 as compared to other fertigation treatments. On the basis of good performance on yield and economics, fertigation with 100% levels of N and K through water-soluble fertilizers can be employed for turmeric transplants.

19.1   INTRODUCTION

Turmeric (*Curcuma longa* L.) is known as the "golden spice" or "spice of life" and it is a herbaceous perennial plant belonging to the family

Zingiberaceae under the order Scitaminae. Turmeric originated in South-East Asia. Its underground modified stem (the rhizomes) are processed and used for various purposes. It is an ancient, most valuable, sacred spice of India containing carbohydrates (69.4%), fiber (2.6%) and appreciable quantity of protein (6.5%) and volatile oil (4.5%) (Manjunath et al., 1991). Curcuminoids in turmeric have anti-inflammatory, antimutagen, anti-cancer, antibacterial, antioxidant, antifungal, antiparasitic, and detoxifying properties (Uechi et al., 2000).

Research and development of newer varieties and improved management techniques are being constantly pursued to enhance the productivity of turmeric. Among the various factors affecting the productivity of turmeric are improper nutritional management practices and inadequate irrigation during critical crop growth stages that can be considered as foremost contributing to low yields. Turmeric is a high input responsive crop. It's extended crop growth period and nutrient exhaustion requires sufficient amount of nutrients and irrigation to produce higher yields with improved quality. Response of turmeric to increased levels of fertilizer has been significant (Parthasarathy et al., 2010). Soil application of fertilizers is associated with nutrient loss through leaching and evaporation. It may also lead to pollution of soil environment.

Turmeric, being a crop with high water requirement, adequate water supply is essential throughout its growth period of 8–9 months. Normally turmeric crop is irrigated at an interval of once in a week depending on soil and weather factors. Increasing scarcity of water often experienced in many parts of turmeric growing regions necessitates alternative means to provide adequate water to the crop without wastage. Drip irrigation has now emerged as one of the innovative approaches to precisely meet the water requirement of many crops. Of late, fertigation (application of fertilizers through drip irrigation) has improved the yield and quality of many horticultural crops (Salo et al., 2002). Turmeric transplants are produced from single bud rhizome. During their growth period, it will require more quantity of nutrients from the external source.

While fertigation can be practiced using conventional fertilizers such as urea and potash with reduced costs, use of water-soluble fertilizers may be effectively employed to improve quality and productivity. Enhancement of yield and quality of various crops have been reported by using water-soluble fertilizers in fertigation (Krishnamoorthy et al., 2015). The influence of water-soluble fertilizers on crop growth and yield of turmeric has not been so far investigated in detail.

With this background in consideration, the present study was taken up to evaluate the effects of different levels of N and K on growth and yield of turmeric var. CO_2.

19.2   MATERIALS AND METHODS

The field experiment to study the effect of fertigation of N and K fertilizers on growth and yield of turmeric transplants (*Curcuma longa* L.) var. CO_2 was carried out at the College Orchard, Department of Spices and Plantation Crops, Horticultural College and Research Institute, Tamil Nadu Agricultural University, Coimbatore during the period from 2014 to 2015. The experiment was laid out in randomized block design, replicated three times. Raised beds of 25 m length, 1 m breadth, 20–25 cm height were formed and turmeric transplants of one-month-old having two fully opened leaves produced from single bud rhizomes were planted in first week of August in paired row system. A spacing of 45 cm between rows within a paired row, 55 cm between two adjacent paired rows and 15 cm within each row was maintained. In treatments receiving fertigation, drip laterals were laid along the length of each paired row at the center with the spacing kept at 1 m between two adjacent laterals. In the control plot, instead of drip laterals, provision for surface irrigation was provided for the paired rows. A venturi assembly was used for mixing fertilizer with irrigation water. Treatments are shown in Table 19.1.

TABLE 19.1   Fertigation Treatments

T_1	Control – 100 % recommended dose of NPK – 150:60:108 kg/ha – through straight fertilizer *i.e.*, Urea & MOP by soil application + surface irrigation
T_2	Fertigation of N+K @ 125 % through straight fertilizers – once in a week
T_3	Fertigation of N+K @ 100 % through straight fertilizers – once in a week
T_4	Fertigation of N+K @ 75 % through straight fertilizers – once in a week
T_5	Fertigation of N+K @ 50 % through straight fertilizers – once in a week
T_6	Fertigation of N+K @ 125 % through water-soluble fertilizers – once in a week
T_7	Fertigation of N+K @ 100 % through water-soluble fertilizers – once in a week
T_8	Fertigation of N+K @ 75 % through water-soluble fertilizers – once in a week
T_9	Fertigation of N+K @ 50 % through water-soluble fertilizers – once in a week

The fertilizers were applied through drip irrigation at weekly intervals by following the schedule by which 40% of total N and 20% of total K were applied from 1[st]to 4th weeks, 10% of total N and 10% of total K were applied from 5th to 8th weeks, 30% of total N and 30% of total K were applied from 9th to 17th weeks. The remaining quantity of 20% N and 40% K were applied from 18th to 34th weeks (Table 19.1). The standard recommended cultural practices (Anonymous, 2013) were followed for managing the crop except for the fertigation treatments envisaged in the study. Data were recorded from the mean of five plants selected randomly from each treatment in each replication on growth and yield parameters (*viz.,* plant height, number of leaves, number of tillers, leaf length, leaf breadth, total dry matter production, yield per plant and yield per plot). The data collected were subjected to statistical analysis using the procedure by Panse and Sukhatme (1985).

19.3   RESULTS AND DISCUSSION

Data regarding the effects of fertigation showed significant variation on the growth and yield components of turmeric transplants in 2014–2015. The highest plant height (140.05 cm), No. of tillers/plant (9.11), No. of leaves/plant (14.83), leaf length (31.35 cm), leaf breadth (10.47 cm) and total dry matter production (31.50 t ha^{-1})were obtained when fertigation of nitrogen and potassium was supplied with 125% of RDF using water-soluble fertilizers. It was followed by performance of fertigation of N+K @ 100% using water-soluble fertilizers – once in a week.

Table 19.2 indicates that with 25% increase of the recommended dose through fertigation, there is an increase in uptake of nitrogen and potassium. As the level of nutrients was increased, there was significant increase in plant height, number of leaves, number of tillers, leaf length and leaf breadth of the turmeric transplants. Increased nitrogen levels had a significant effect on the height of turmeric plant. Plant height is closely related to the development of leaf area, which indicates that the height of the plant supports the development of leaf area. NPK application will encourage the growth of organs associated with photosynthesis of leaves. The increased height of the aerial shoot may be helpful for better exposure of the leaves to the sun thereby increasing the photosynthetic efficiency. Increase in plant height with optimum fertigation levels have been reported in other crops such as onion and paprika (Muralikrishnasamy et al., 2005; Prabhu, 2006).

TABLE 19.2 Influence of Straight and Water-Soluble Fertilizers on Growth Characters of Turmeric Transplants at 210 DAP (Days After Planting)

Treatments	Plant height	No. of leaves	No. of tillers	Leaf length	Leaf breadth	Total dry matter production
	cm	-	-		cm	t ha^{-1}
T$_1$	113.00	11.72	4.88	28.62	9.26	27.12
T$_2$	131.31	13.40	8.24	31.16	10.42	29.89
T$_3$	127.52	12.73	7.42	30.28	10.22	28.42
T$_4$	121.22	12.60	5.60	29.41	9.69	28.14
T$_5$	114.48	11.90	5.10	29.16	9.45	26.54
T$_6$	**140.05**	**14.83**	**9.11**	**31.35**	**10.50**	**31.50**
T$_7$	133.31	14.01	8.25	31.21	10.44	30.74
T$_8$	127.53	13.33	7.83	30.66	10.34	29.12
T$_9$	125.73	12.67	5.65	29.73	9.81	27.87
Mean	**126.02**	**13.02**	**6.90**	**30.18**	**10.01**	**28.82**
SEd	**2.6524**	**0.2231**	**0.1789**	**0.6294**	**0.2344**	**0.653**
CD (P = 0.05)	**5.6229**	**0.4729**	**0.3793**	**1.3344**	**0.4969**	**1.384**

In general, higher levels of N, P, and K enhanced the production of leaves. Enhanced vegetative growth as a result of higher levels of fertilizers was reported in ginger grown under artificial shade (Ancy and Jayachandran, 1996). The lowest number of tillers per plant was produced at the soil application of fertilizers, indicating that frequent application of nitrogen and phosphorus played more important roles in increasing production of tillers. This may be attributed to the rapid conversion of synthesized carbohydrates into protein and consequently the increase in number and size of growing cells, resulting ultimately in increased number of tillers (Agarwal and Singh, 2009). Aulakh and Malhi (2005) advocated increased response of the applied nitrogen with increase in potassium levels contributing to the improvement in crop growth. Higher doses of nitrogen played an important role in synthesis of protein which is important for buildup of new cells and consequently influenced the growth (Satyareddi and Angadi, 2014).

An increase in total dry matter production of the plant was apparent due to fertigation treatments compared to soil application of fertilizers. In the treatment T$_6$ (Fertigation of N+K @ 125% through water-soluble fertilizers – once in a week), significantly higher total dry matter production of the plant was registered at different stages of observation. This was due to higher underground rhizome mass. The highest uptake of nutrients could have led to maximum dry matter accumulation (Somasundaram and Shanthi, 2014).

Table 19.3 indicates the data on yield. Yield response to fertigation was also significantly higher in the treatment T_6 (Fertigation of N+K @ 125% through water-soluble fertilizers – once in a week) as exhibited by higher fresh rhizome yield/plant (401.00 g), yield/plot (109.09 kg/25 m²), estimated fresh rhizome yield/ha (43.64 t ha⁻¹) and estimated cured rhizome yield/ha (7.92 t ha⁻¹). This was closely followed by T_7, which recorded a yield of 396.00 g/plant, 109.06 kg/plot (25 m²), 43.62 t/ha of estimated yield and an estimated cured rhizome yield of 7.90 t/ha, respectively. Yield is a complex character and associated with several yield contributing traits. Fertigation using water-soluble fertilizers at 125 and 100% recommended levels, significantly, and consistently proved better for these parameters followed by fertigation using straight fertilizers at 125% recommended levels. In case of soluble fertilizers, the nutrients become available readily throughout the growth stages of crop to produce optimum yield. However,

TABLE 19.3   Influence of Straight and Water-Soluble Fertilizers on Rhizome Yield of Turmeric Transplants

Treatment	Yield per plant	Yield per plot (25 m²)	Estimated fresh rhizome yield	Estimated cured rhizome yield	B:C ratio
	g	kg	t ha⁻¹		-
T_1	317.00	85.35	34.14	5.80	2.99
T_2	389.00	99.32	39.73	6.89	3.50
T_3	374.00	93.06	37.22	6.61	3.40
T_4	350.00	87.48	34.99	6.18	3.28
T_5	329.00	86.57	34.63	5.71	2.98
T_6	**401.00**	**105.65**	**42.26**	**7.92**	**3.56**
T_7	396.00	100.52	40.21	7.90	3.65
T_8	381.00	95.02	38.01	6.75	3.19
T_9	365.00	89.62	35.85	6.35	3.16
Mean	**366.89**	**93.62**	**37.45**	**6.68**	
SEd	**7.0746**	**1.8321**	**0.6486**	**0.166**	
CD (P = 0.05)	**14.9976**	**3.8839**	**1.3749**	**0.352**	

straight fertilizers when applied into soil they may get leached out, volatilize or get fixed into the soil and hence they become unavailable to crop for their growth and development and hence crop does not produce optimum yield with its full potential. Similar results were also reported by Ughade and Mahadkar (2015).

19.4 SUMMARY

Balanced use of fertilizers will improve turmeric yield and it contributes greatly to the economic viability of the crop. The assessment of economics of cultivation due to different fertigation treatments clearly indicated the superiority of fertigation treatments over the conventional treatment. The results revealed that the benefit-cost ratio was higher under fertigation with water-soluble fertilizers compared with that of fertigation with straight fertilizers. The highest benefit-cost ratio was recorded with the fertigation of N+K @ 100% through water-soluble fertilizers – once in a week (T_7) and recorded the benefit-cost ratio of 3.65 compared to other fertigation treatments. The results in this study showed that fertigation of N+K @ 100% through water-soluble fertilizers applied at weekly interval can enhance productivity and help to gain higher returns in turmeric transplants.

KEYWORDS

- *Curcuma longa* L.
- drip irrigation
- fertigation
- turmeric

REFERENCES

Agarwal, S. K., & Singh, R., (2009). Growth and yield of turmeric (*Curcuma longa*) as influenced by levels of farm yard manure and nitrogen. *Indian J. Agron., 46*(3), 462–467.

Ancy, J., & Jayachandran, B. K., (1996). Nutrient requirement of ginger (*Zingiberofficinale* R.) under shade. *Indian Cocoa, Arecanut, and Spices J., 20*(1), 115–116.

Anonymous, (2013). *Crop Production Techniques of Horticultural Crops.* Horticultural College and Research Institute, Tamil Nadu Agricultural University, Coimbatore, Tamil Nadu, p. 20.

Aulakh, M. S., & Malhi, S. S., (2005). Interactions of nitrogen with other nutrients and water: Effect in crop yield and quality, nutrient use efficiency, carbon sequestration and environmental pollution. *Adv. Agron., 86*(1), 341–409.

Krishnamoorthy, C., Soorianathasundaram, K., & Mekala, S., (2015). Effect of fertigation on FUE, quality, and economics of cultivation in turmeric (*Curcuma longa* L.) cv. BSR–2. *Int. J. Agric. Sci. Res., 5*(1), 67–72.

Manjunath, M. N., Sattigeri, V. V., & Nagaraj, K. V., (1999). Curcumin in turmeric. *Spice India, 4*(3), 7–9.

Muralikrishnasamy, S., Veerabadran, V., Krishnasamy, S., Kumar, V., & Sakthivela, S., **(2005)**. Micro sprinkler irrigation and fertigation (*Allium cepa*). In: *7th International Micro Irrigation Congress* (p. 49–57). Kuala Lumpur, Malaysia.

Panse, V. G., & Sukhatme, P. V., (1985). *Statistical Methods for Agricultural Workers.* Indian Council of Agricultural Research, New Delhi, p. 112.

Parthasarathy, V. A., Dinesh, R., Srinivasan, V., & Hamza, S., (2010). Integrated nutrient management in major spices. *Indian J. Fert., 6*(1), 110–128.

Prabhu, T., (2006). *Standardization of Fertigation Techniques in Paprika (Capsicum Annuum var. Longum L.) Under Open and Coconut Shade Conditions.* PhD (Hort.) Thesis for Tamil Nadu Agricultural University, Coimbatore, Tamil Nadu, p. 230.

Salo, T. T., Suojala, T., & Kallela, M., (2002). The effect of fertigation on yield and nutrient uptake of cabbage, carrot, and onions. *Acta Hort., 571*, 235–241.

Satyareddi, S. A., & Angadi, S. S., (2014). Response of turmeric (*Curcuma longa* L.) varieties to irrigation methods and graded levels of fertilizer. *Res. Environ. Life Sci., 7*(4), 237–242.

Somasundaram, E., & Shanthi, G., (2014). Sustainable production packages for turmeric. *Building Organic Bridges, 2*(1), 615–618.

Uechi, S., Miyagi, Y., Ishimine, Y., & Hongo, F., (2000). Antibacterial activity of essential oils from *Curcuma* sp. (*Zingiberaceae*) cultivated in Okinawa against foodborne pathogenic bacteria. *Japan J. Trop. Agric., 44*(1), 138–140.

Ughade, S. R., & Mahadkar, U. V., (2015). Effect of different planting density, irrigation, and fertigation levels on growth and yield of brinjal (*Solanum melongena* L.). *The Bioscan, 10*(3), 1205–1211.

CHAPTER 20

PERFORMANCE OF DRIP-FERTIGATED TUBEROSE (*POLIANTHES TUBEROSA* L.) UNDER POLYETHYLENE (PE) MULCHING

J. KABARIEL, M. KANNAN, and M. JAWAHARLAL

ABSTRACT

The field experiment was conducted to standardize the effects of irrigation regimes and fertigation levels on certain yield parameters of tuberose cv. Prajwal. The experiment was conducted in fully randomized block design (FRBD) with two factors: Irrigation (75% WR_c, 100% WR_c, 125% WR_c) and fertigation levels (100% Water-soluble fertilizers (WSF), 100% Straight fertilizers (SF), 75% WSF + 25% SF, 50% WSF + 50% SF, 25% WSF + 75% SF, 100% WSF without mulch and 100% SF without mulch). The experiment was conducted during June 2013 to May 2014. Irrigation was done through drip system once in two days, based on plant water requirement by pan evaporation method. Fertigation was done once in a week. The fertilizer dose was based on the requirement of the crop growth stage. Observations were recorded for different yield parameters of tuberose. Among these parameters: days taken for spike emergence, flowering duration, number of spikes per clump, spike length and rachis length proved their superiority in the treatment, which received irrigation 125% WR_c + 100% water-soluble fertilizers under mulching.

20.1 INTRODUCTION

Tuberose (*Polianthes tuberosa* L.) is grown commercially in India mainly due to its color, elegance, and fragrance. Among different flowers grown in India, tuberose has attained prime position because of its popularity as cut flower, loose flower and for its potential in the perfume industry. Tuberose has high demand in the market and its production is highly profitable. Irrigation and nutrients availability largely affects the growth of tuberose production. Fertigation is the most advanced and efficient practice of fertilization to take advantage of water and nutrients. The right combination of water and nutrients is a key for high growth of tuberose.

The present investigation was undertaken to study the effects of drip irrigation and fertigation on yield parameters of tuberose.

20.2   MATERIALS AND METHODS

The experiment was conducted at the Botanical Garden, Tamil Nadu Agricultural University, Coimbatore during 2013–2014, in a sandy clay loam soil with low available N (188 kg ha^{-1}), high P (47 kg ha^{-1}) and high K (724 kg ha^{-1}), pH 8.87 with 1.01 EC (dS m^{-1}). Uniform sized bulbs were planted in black polythene mulched raised beds in paired row system of planting at a spacing of 150 x 45 x 45 cm. There were two main factors:

Irrigation levels:

I_1 75%WR$_c$ I_2 100%WR$_c$ I_3 125%WR$_c$

Fertigation levels:

F_1 100% Water-soluble fertilizers (WSF),
F_2 100% Straight fertilizers (SF),
F_3 75% WSF + 25% SF,
F_4 50% WSF + 50% SF,
F_5 25% WSF + 75% SF,
F_6 100% WSF without mulch, and
F_7 100% SF without mulch.

In total, there were 21 treatment combinations, which were replicated thrice. The 75% recommended dose of phosphorus was applied in the form of a single super phosphate as basal application. Irrigation was provided once every two days based on the pan evaporation formula:

$$WRc = \{[CPE \times Kpx\ A \times Kc \times Wp] - Re\} \qquad (1)$$

where, WRc = Water requirement (liters per plant); CPE = Cumulative pan evaporation (mm); Kp = Crop factor (0.8); Kc = Crop coefficient (Initial– 0.7; Mid–1.05; and End of crop- 0.80); Wp = Wetting percentage (80%); A = Area per plant (m^2); and Re = Effective rainfall (mm).

The water-soluble fertilizers (WSF) in this study were Poly feed (19:19:19), KNO_3 (13:0:45), Urea (46%N), SSP (16%P) and MOP (60%K). The crop growth was classified into bulb planting to establishment (2 weeks), vegetative (10 weeks) and spike emergence to flowering stage (40 weeks). Therefore based on the growth stage, the percent requirement of fertilizers was calculated and applied through fertigation once each week. Yield parameters of tuberose were evaluated.

20.3 RESULTS AND DISCUSSION

The aim of any floriculture research is to get good quality of flowers in terms of increased flower shoot length and number of flower buds per unit area. The yield attributing characters were: days taken for first spike emergence (75.66 days), flowering duration (22.06 days), number of spikes clump^{-1} (4.13), spike length (99.62 cm) and rachis length (23.46 cm); and these were significantly influenced by different irrigation regimes and fertigation levels (Tables 20.1–20.5).

TABLE 20.1 Effect of Drip Irrigation Regimes and Fertigation Levels on Days Taken for Spike Emergence in Tuberose

	F_1	F_2	F_3	F_4	F_5	F_6	F_7	**Mean**
I_1	88.88	94.11	89.33	92.55	94.66	92.22	127.00	**96.96**
I_2	85.88	87.11	86.33	87.44	88.44	87.33	88.78	**87.33**
I_3	75.66	81.89	77.88	80.11z	82.55	79.33	84.55	**80.28**
Mean	**83.47**	**87.70**	**84.51**	**86.70**	**88.55**	**86.29**	**100.11**	
	SED	**CD(0.05)**		**SED**	**CD(0.05)**		**SED**	**CD(0.05)**
I	0.312	0.631	F	0.382	0.964	IF	0.826	1.671

Days to first spike appearance are an important character, which decides the early yield (precocity) of the crop. Early flowering might be due to the combined effect of fertigation along with drip irrigation creating a conducive source-sink relationship. The commencement of early flowering noticed in I_3F_1 with drip irrigation at 125%WR_c + 100% WSF with mulch. This might be due to the action of WSFs, which are easily translocated by higher moisture regimes. Nutrient is a constituent of proteins, amino acids, nucleic acid, various enzymes, and coenzymes, which are associated with the increased leaf length and leaf area that significantly contribute to higher photosynthesis and thereby increased transformation of manufactured food material from source (leaf) to sink (spike). The results were in conformity with findings by Beniwal et al. (2005) in chrysanthemum. Similar results were also obtained by Marchner (1983), Potti and Arora (1986) in chrysanthemum, Khimani (1991) in blanket flower, Mukherjee et al. (1994), Sharma and Singh (2001) in gladiolus, Krishna et al. (1999) in carnation, and Selvaraj (2007) in *J. grandiflorum*.

TABLE 20.2 Effect of Drip Irrigation Regimes and Fertigation Levels on Flowering Duration of Tuberose

	F_1	F_2	F_3	F_4	F_5	F_6	F_7	Mean
I_1	17.40	13.26	16.73	15.40	12.86	16.00	12.60	**14.89**
I_2	18.86	18.00	18.66	18.20	17.86	18.36	17.80	**18.25**
I_3	22.06	18.86	21.40	20.33	18.53	21.00	19.20	**20.20**
Mean	**19.44**	**16.71**	**18.93**	**17.97**	**16.42**	**18.45**	**16.53**	
	SED	CD (0.05)		SED	CD (0.05)		SED	CD (0.05)
I	0.151	0.307	F	0.232	0.469	IF	0.401	0.812

TABLE 20.3 Effect of Drip Irrigation Regimes and Fertigation Levels on Number of Spikes per Clump of Tuberose

	F_1	F_2	F_3	F_4	F_5	F_6	F_7	Mean
I_1	1.86	1.06	1.53	1.40	0.73	1.46	0.46	**1.21**
I_2	2.46	2.26	2.40	2.26	2.13	2.33	2.13	**2.28**
I_3	4.13	3.26	3.86	3.66	3.20	3.80	3.13	**3.58**
Mean	**2.82**	**2.20**	**2.60**	**2.44**	**2.02**	**2.53**	**1.91**	
	SED	CD (0.05)		SED	CD (0.05)		SED	CD (0.05)
I	0.035	0.072	F	0.054	0.111	IF	0.095	0.192

Spike length, rachis length and number of spikes per clump are important factors, which influence the size and quality of the florets. The results of the study indicated that increase in spike length, rachis length and number of spikes per clump in the plants applied with 125%WR$_c$ drip irrigation + 100% WSF with mulch might be due to better uptake of nutrients especially N, due to the combination of NO_3^- and NH_4^+ as water-soluble fertilizers. Further, the presence of NO_3^- with negative ions and all the positive ions of K^+, which was fed through multi-K resulted in increased number of spikes with higher spike length and rachis length. These results are in agreement with the reports by Vaugham et al. (1985), Ashok Kumar (2006) in paprika, and Sathish (2006) in turmeric.

TABLE 20.4 Effect of Drip Irrigation Regimes and Fertigation Levels on Spike Length (cm) of Tuberose

	F_1	F_2	F_3	F_4	F_5	F_6	F_7	Mean
I_1	82.70	76.72	80.00	77.19	75.73	79.62	70.12	**77.44**
I_2	93.01	88.35	92.62	89.64	85.40	90.94	84.55	**89.21**
I_3	99.62	94.93	96.17	95.37	94.23	96.55	93.44	**95.83**
Mean	**91.77**	**86.66**	**89.77**	**87.40**	**85.12**	**89.03**	82.70	
	SED	CD(0.05)		SED	CD(0.05)		SED	CD(0.05)
I	0.715	1.446	F	1.093	2.209	IF	1.620	3.242

TABLE 20.5 Effect of Drip Irrigation Regimes and Fertigation Levels on Rachis Length (cm) of Tuberose

	F_1	F_2	F_3	F_4	F_5	F_6	F_7	Mean
I_1	16.86	15.42	16.52	15.82	15.01	16.01	14.92	**15.79**
I_2	18.52	17.47	18.11	17.82	17.21	18.04	17.03	**17.74**
I_3	23.46	19.43	20.96	19.98	19.06	20.42	18.98	**20.32**
Mean	**19.61**	**17.44**	**18.53**	**17.87**	**17.09**	**18.15**	16.97	
	SED	CD(0.05)		SED	CD(0.05)		SED	CD(0.05)
I	0.715	1.446	F	1.093	2.209	IF	1.620	3.242

The results of the present study suggested that higher level of irrigation water (125%) and 100% water-soluble fertilizers through drip fertigation system under polythene mulching significantly increased the days taken for spike emergence, flowering duration, number of spikes per clump, spike length and rachis length. Fertilizers applied through drip irrigation at frequent intervals in small quantities increased the fertilizer use efficiency and nutrient uptake as it prevents the loss of nutrients by leaching, erosion as well as by weeds. Effective and efficient utilization of water and nutrients by the plants resulted in better plant growth and volume.

20.4 SUMMARY

The present investigation focuses on the effects of drip irrigation and fertigation on yield parameters of tuberose under mulching. Among yield parameters (such as: days taken for spike emergence, flowering duration, number of spikes per clump, spike length and rachis length) showed their superiority in the treatment, which received irrigation at 125% WR_c + 100% WSFs under mulching.

KEYWORDS

- fertigation
- irrigation
- pan evaporation
- spike length
- tuberose

REFERENCES

Beniwal, B. S., Ahlawat, V. P., & Singh, S., (2005). Effect of nitrogen and phosphorus levels on flowering and yield of chrysanthemum (*Chrysanthemum morifolium*) cv. Flirt. *Crop Res., 30*(2), 177–180.

Khimani, R. A., (1991). *Standardization of Production Technology in Gaillardia (Gaillardia Pulchella Foug).* PhD Thesis submitted to University of Agricultural Sciences, Dharwad, p. 199.

Marchner, H., (1983). Introduction to the mineral nutrition of plants. In: *Handbook Plant Physiology* (Vol. 154, pp. 31–38).

Marchner, H., (1995). *Mineral Nutrition of Higher Plants.* 2nd edn., Special Publications of the Society for General Microbiology, Academic Press, New York, p. 889.

Potti, S. K., & Arora, J. S., (1986). Nutritional studies in gladiolus. *Punjab Hort. J., 26* (1–4), 125–128.

Satish, G., (2006). *Studies on Effect of Bioregulants on Yield and Quality of Turmeric (Cutcuma Longa) var. BSR–2*. M.Sc., (Hort.) Thesis submitted to Tamil Nadu Agricultural University, Coimbatore, p. 125.

Selvaraj, V., (2007). *Standardization of Irrigation and Fertigation Techniques in Jasmine (Jasminum Grandiflorum)* var CO_2. PhD (Hort.) Thesis submitted to Tamil Nadu Agricultural University, Coimbatore, p. 245.

INDEX

9 781774 634677